# PETROLEUM:

A

# HISTORY OF THE OIL REGION

OF

## VENANGO COUNTY, PENNSYLVANIA.

ITS RESOURCES, MODE OF DEVELOPMENT, AND VALUE:
EMBRACING A DISCUSSION OF ANCIENT OIL OPERATIONS;
WITH A MAP, AND ILLUSTRATIONS OF OIL
SCENES AND BORING IMPLEMENTS.

BY
REV. S. J. M. EATON,
PASTOR OF THE PRESBYTERIAN CHURCH, FRANKLIN, PA.

"The rock poured me out rivers of oil."

PHILADELPHIA:
J. P. SKELLY & CO.,
732 CHESTNUT STREET.
1866.

STEREOTYPED AND PRINTED BY
ALFRED MARTIEN

# PREFACE.

---

THIS work aims at nothing more than a mere popular description of the oil region, of the early history of petroleum operations, of the manner of boring oil wells, bringing the oil to the surface, and preparing it for use. As this region is attracting a large share of public attention it has been thought proper to devote a few pages to its ancient history; by way of illustrating its changes, a brief chapter is also added, bringing down its history to the period when modern oil operations commenced.

As there is much curiosity abroad as to the manner of boring deep wells in the rock, and pumping from great depths, these processes are detailed with great minuteness and particularity.

The chapter on the origin of petroleum, although claiming little originality, is presented with much diffidence. From the circumstances of the case, we are not likely soon to pass beyond mere theory in the matter, and, with the light we yet have, that presented seems the most plausible and consistent with the facts exhibited in boring and pumping.

It was manifestly injudicious to attempt to give the number and production of each well in given localities, for what would be true to-day would be unreliable next week, as the development in new wells is constantly changing the features of particular localities. Consequently a general view is given, in order to arrive at general results.

The design of the work is set forth on its title page—a history of the oil region of Venango county, Pennsylvania. It is not by this designed to ignore the existence of petroleum in neighboring counties, in this or in other States; this is admitted, and it is also anticipated that other regions still may and will develop large resources of oil; the design is to discuss the matter in its bearings on this region, where we find the evidences of its earliest development, and where, as yet, its largest resources have been brought to light.

Grateful acknowledgments are due to gentlemen who have aided in collecting materials for this work, and thanks are hereby given.

The Oil scenes in this volume are from photographs by A. D. Deming, an enterprising artist of Oil City, Pa.

FRANKLIN, Pa., November, 1865.

# CONTENTS.

CHAPTER XIII.—FRANKLIN, FRENCH CREEK, AND SUGAR CREEK.



CHAPTER XIV.—OIL CREEK.



CHAPTER XV.—ALLEGHENY RIVER TERRITORY.



CHAPTER XVI.—USES OF PETROLEUM.

# PETROLEUM.

## CHAPTER I.

### ANCIENT HISTORY OF VENANGO COUNTY.

SUDDEN and unexpected changes are characteristic of the age in which we live. Progress and development seem remarkably animated and energetic during this latter half of the nineteenth century. The sands seem to run rapidly through the glass, as we approach the grand climacteric in the world's history. The Almighty God appears to be opening his great treasure house to provide for the grand and important changes, that in his wise and inscrutable Providence, he is working among us here.

Venango county, Pennsylvania, has suddenly attained to an interest, unsurpassed by any other region of the United States. From being one of the most unproductive and obscure, it has become the first in interest, and wealth and influence among its sister counties, in all the broad range from the Atlantic to the Pacific. Some facts and brief notes, therefore, will be of interest in regard to its ancient history. The landmarks are becoming indistinct, and the light fading, that years ago, would have thrown a wondrous interest around this now popular region. There are many circumstances that lead to the belief that this portion of the valley of the Allegheny has been the scene of strange and wonderful incidents in years long gone by. There are footprints yet remaining of a strange and mysterious people, that dwelt there, and labored there, long before the advent of the Indian tribes. But a few imperfect footprints are all that remain. Their broad pathway is almost obliterated by the fast accumulating sands of time. Tradition is voiceless; history has made no record; and we can but wish for the unwritten annals of the past.

In regard to the possession of this region by the Indians, it is probable, from what feeble light we have, that at a very early day, about the close of the fifteenth century, the whole country extending along the extreme portions of the shore of Lake Erie, and for some distance southward, was in possession of a powerful tribe known as the Eries, and from which it is probable the lake itself takes its name. Tradition relates that this tribe was peacefully disposed, and at one time was ruled by a queen named Yagowannea; who, like the queen of Palmyra, after ruling with dignity and justice, at length fell a victim to the jealousy and intolerance of the surrounding tribes.

Schoolcraft relates that the Eries being pressed by their enemies, gradually moved towards the Ohio, or Allegheny river, where soon after their council fires were put out, and they ceased to be known as a tribe. The date of this extinction was 1653.

That terrible confederacy of the Indian tribes, known first as the Five, and afterwards as the "Six Nations," was formed and extended their conquests far and wide. From the Mohawk river, they extended their sway westward like the ocean tide, sweeping everything from their pathway until they reached the Mississippi, kindled their council fires throughout the whole northern portion of the Allegheny valley. The Eries were exterminated, in this path of conquest, as a nation, and their broken fragments absorbed by these "Romans of America."

At the time when the Oil regions first became known to civilized men, the "Six Nations," or, as they were called by the French, "Iroquois," were in possession of the land. During the long wars, that continued for nearly a century between England and France, these people were generally on the side of England; and during the troubles of the Revolution, their influence was as a general thing with the British. But they too, soon began to follow in the wake of the Eries. They are in the sear and yellow leaf. As civilization advanced, their strength passed away, and now, but a few scattered fragments remain, one of which holds a limited possession in the northern outskirts of the oil region, on the upper Allegheny.

The remains of antiquity that are still found in this portion of the Allegheny valley, satisfy us, that it was considered of no small importance in the history of the past. Franklin, the county seat of Venango, is situated in a flexure of the Allegheny, and at the junction of French creek with that river. It is surrounded on all sides by bold, precipitous hills, rising to the height of some six hundred feet. On the highest crest of land, on the Northeastern side of the town, is a remarkable ancient work, that seems evidently designed as a kind of "lookout" over the river and creek. From this point, there is an uninterrupted view of the river for several miles upward and downward, as well as of French creek, with all its sinuosities. This work has something the appearance of the modern "rifle pits." It is in the form of an inverted cone, about eight feet in diameter, at the top, and at the present time about six feet in depth; rudely walled up to the surface with rough masonry. It is situated at the precise point, most favorable to secure a view of the valley in which the town is situated, as well as of the creek and river above and below. It is said that two pits similar to this, are to be seen, one on the eastern side of the river, and the other on the southwestern side of the town; the three forming a triangle, and together commanding the approaches to the town, in every possible direction. Whether these are the work of the French, during their terrible struggle to plant and retain their banner all over this country; or are due to the unhistorical race, some evidence of whose sojourn in this valley, will be recorded in a succeeding chapter, cannot now be ascertained. The more probable opinion, however, is that their date is anterior to that of the French possession.

About six miles south of Franklin, and nine by the course of the river, is a rock, a noted landmark in Indian history. It has long been known to the present inhabitants as "The Indian god." It is on the eastern margin of the river, and during freshets is completely submerged. It is an immense boulder in a deflection of the river, standing at an inclination of about fifty degrees to the horizon, and is about twenty-two feet in length by fourteen in breadth, with an inscription on its inclined face. As to this inscription and its interpretation, we cannot do better than quote from Schoolcraft's work on the Indian Tribes.

"The inscription itself appears distinctly to record, in symbols, the triumphs in hunting and war. The bent bow and arrow are twice distinctly repeated. The arrow by itself is repeated several times, which denotes a date before the introduction of fire-arms. The animals captured, to which attention is directed by the Indian pictographist, are not diet, or common game, but objects of higher triumph. There are two large panthers, or cougars, variously depicted; the lower one in the inscription denoting the influence, agreeably to pictographs heretofore published, of medical magic. The figure of a female denotes, without a doubt, a captive, various circles representing human heads denote deaths. One of the subordinate figures depicts, by his gorgets, a chief. The symbolic sign of the raised hand, drawn before a person represented with a bird's head, denotes, apparently, the name of an individual or tribe."

At the foot of the large rock is a smaller one, containing a single figure.

About the close of the *fifteenth* century, the struggle commenced between England and France, for the possession of the Allegheny and Mississippi valleys. England's claim was set forth in the following language:—

"That all the lands, or countries, westward from the Atlantic Ocean to the South Sea, between 48° and 34° of north latitude, were expressly included in the grant of King James the First, to divers of his subjects, so long since as the year 1606, and afterwards confirmed in 1620."* The original ground of this claim was, in

*Colonial Records of Pennsylvania.

part, original settlement, but chiefly treaties of purchase from that wonderful confederacy, the "Six Nations," that was found in possession. The French based their claim upon original discovery by Marquette and La Salle, together with their construction of their treaties of Ryswick, Utrecht, and Aix la Chapelle. But this claim was based upon very insecure grounds, and seems rather prompted by ambition and lust of power. They had at this time possessions in Canada on the North, and at the mouth of the Mississippi on the South, and wished to unite these possessions and thus present a barrier to the rapidly advancing tide of power that was sweeping westward under the auspices of England. As early as the beginning of the eighteenth century, Bancroft tells us, that "not a fountain bubbled on the west of the Allegheny, but was claimed as being within the French empire."

In the summer of 1749, Gallisoniere, then Governor of Canada, sent Louis Celeron with a party to place leaden plates along the whole line, extending from Lake Erie to the mouth of the Mississippi, as evidences of the French claim to the territory.* One of these was found buried at Franklin, Pa., bearing the following inscription:—"In the year 1749, reign of Louis XV, King of France, M. Celeron, commandant of a detachment by Monsieur the Marquis of Gallisoniere, commander-in-chief of New France, to establish tranquillity in certain Indian villages of these cantons, has buried this plate at the confluence of the Toradakoin, this 20th July, near the river Ohio, otherwise Beautiful river, as a monument of renewal of possession which we have

*Western Annals.

taken of said river, and all its tributaries; and of all lands on both sides as far as the sources of said rivers; inasmuch as the preceding kings of France have enjoyed it by their arms and by treaties; especially by those of Ryswick, Utrecht, and Aix la Chapelle."

By the Ohio, we are of course to understand the Allegheny, and by the Toradakoin, French creek.

In pursuance of the determination to hold possession of this territory, the French proceeded to erect a chain of forts, extending from Presque Isle, or Erie, to the Ohio river. Fort Venango, one of this chain, was commenced in 1753, and completed in April, 1754. We have an interesting account of the erection of this Fort in a deposition made by Stephen Coffin, before Col. Johnston of New York. Coffin had been a prisoner among the French, and accompanied them on this fort building expedition. He says that the detachment consisted of three hundred men, and left Canada in January, 1753. After describing the landing at Erie and the erection of "Fort Le Presque Isle," at that place, he continues:—"As soon as the fort was finished, we marched southward, cutting a wagon road through a fine level country, twenty-one miles to the River aux Boeufs" (now French creek, at Waterford). "We fell to work cutting timber, boards, &c., for another fort, while Mr. Morang ordered Monsieur Bito, with fifty men, to a place called by the Indians Ga-na-ga-rah-ha-re (now Franklin) on the banks of Belle Riviere, (now Allegheny river) where the river Boeufs (now French creek,) empties into it. In the meantime Morang had ninety large boats or batteaux made to carry down the baggage, provisions, &c., to it. Monsieur Bito, on coming to said Indian place, was asked what he wanted or intended. Upon answering, that it was the intention of their father, the Governor of Canada, to build a trading house, for their and all their brothers' convenience, he was told that the lands were theirs, and that they would not have them build upon it."

Thus far Coffin's statement goes; but it appears that the scruples of the Indians were overcome, and the fort known as Venango was commenced late in the year 1753.

In November of the same year, George Washington was commissioned by Governor Dinwiddie to proceed from Fort Pitt, up the Allegheny to Fort Venango and Le Boeufs, to make observations in regard to the French occupation, forts, number of men, and probable intentions. In pursuance of this mission, Mr. Washington, then in his twenty-second year, arrived at Venango on the 4th of December. The following extracts from his journal will give some idea of his movements.

"This is an old Indian town, situated at the mouth of French creek on the Ohio."

"We found the French colors hoisted at a house from which they had driven Mr. John Frazier, an English subject." After obtaining an interview with three officers from the fort, and the wine, or more probably poor whisky, having made them social and communicative, the journal proceeds:—

"They told me that it was their absolute design to take possession of the Ohio, and by G——, they would do it; for although they were sensible that the English could raise two men for their one, yet they knew their motions were too slow and dilatory to prevent any undertaking of theirs."

Washington and his little party appear to have

crossed in the vicinity of the upper bridge, and proceeded to Fort Le Boeufs or Waterford.

We next find Fort Venango spoken of in 1759, in a letter from Colonel Mercer, English commander at Fort Pitt. Colonel Mercer had sent two Indian scouts up to Fort Venango, who reported that there was then at the fort seven hundred French, and one thousand Indians, which force was destined to attack Fort Pitt.

In this year (1759) the French finally withdrew their forces to strengthen Fort Niagara, and untoward events causing them to abandon their efforts towards holding the country, they never afterwards undertook the possession of Fort Venango. This fort was situated on the bank of the Allegheny, about thirty rods below the present Allegheny bridge. A draught, said to be the original drawing by the French engineer, is still in the possession of Wm. Reynolds of Meadville, Pa.

There are the ruins of another fort near the Allegheny bridge, that the old settlers were in the habit of calling the English Fort. There appears to be considerable confusion of ideas, in regard to this fort. That it was in possession of the English after the country was abandoned by the French, is quite likely; but the probabilities are very strong, that it, too, was built by the French, after the destruction of the first one. In 1763, while Fort Venango was in possession of the English garrison, a large party of Senecas, on pretence of friendship, gained admission, massacred the garrison, tortured Lieutenant Gordon, the commander, over a slow fire, and finally burned the fort.

Opposite Fort Venango, and on the eastern side of the river, Henry Decourcy affirms on the authority of an old map preserved in Quebec, that Fort Michault was built about the same time as Venango. In 1757, Monsieur Chauvignerie, Jr., a French prisoner, testified before a justice of the peace to the same effect:—"My father was a lieutenant of marines, and commandant of Fort Michault, built lately at Venango." Of such a fort, the oldest settlers can give no information, nor are there any traces of ruins that can be identified on that side of the river. It is possible, however, that this may be identical with what is now known as the English fort, erroneously placed on the Quebec map on the opposite side of the river.

In the neighborhood of these forts, some very fine species of grape were found by the first settlers, and have been propagated to the present time. One of these varieties is known as the "Venango" grape. The cuttings were no doubt originally brought from La belle France.

We next find a more peaceful mission organized to visit Venango county. It was in the fall of 1767. A Moravian missionary, the Rev. David Zeisberger, accompanied by two Christian Indians as assistants, succeeded in penetrating the wilderness, and reaching a point on the Allegheny near the mouth of Tionesta creek, and about thirty miles above Franklin. This good brother appeared among the red men of the forest unarmed, plainly dressed, and simple in manners, with the one avowed object, that of doing them good. At first the missionary was regarded with suspicion, but gradually won his way into the favor of the people, until a chapel was built, corn planted, and the work commenced. The place was called by the Indians "Goschgosking" or according to another authority "Goshgoshunk." An Indian prophet, named Wangomen, declaimed against the new religion. Like Demetrius of old, when Paul preached against idolatry, he saw their craft was in danger, so that no small stir was excited. The prophet declared that the Great Spirit was angry at the innovation, and was blighting the corn and driving away the game, as a punishment for their infidelity. The place soon became too uncomfortable for the good brethren, when they sought a new location for their mission, about fifteen miles farther up the river, at a town called Lawunakhannak, probably in the neighborhood of Hickorytown. But the troubles that began at the former point followed them to the new location. The religious teachers and medicine men looked upon the strangers as their rivals, jealousies were fomented, until in the month of April, 1770, the heroic little band of missionaries were obliged to break up their mission, and with sorrowful hearts launch their canoes and set out for a more propitious field. This mission appears to be the first effort to plant the gospel in what is now known as the oil region of Venango county.

A word in regard to the names used in the ancient history of the county, is necessary in this place to avoid confusion. The Indians and French alike considered what are now known as the Allegheny and Ohio as one and the same river. In fact the names signify the same, in different dialects of the Indian country. Allegheny is from the Delaware language, and O-hee-o from the Seneca, both meaning beautiful water. Hence the French term "La Belle Riviere" or Beautiful river. In the earlier periods of the history of the country it was known as the Ohio. What is now known as French creek appears to have been known by the Indians as To-ra-da-koin. By the French it was first called Riviere aux Boeufs, or Buffalo river, afterwards Venango river, a corruption probably of In-nan-ga-eh, in the Seneca language referring to a rude figure carved on a tree, when first discovered by this tribe. By George Washington it was re-christened, at the time of his visit in 1753, French creek, which name it still bears. The change conveys to us something of an idea of the practical nature of Washington's mind, in rejecting the beautiful flowing Indian name for the common-place modern one.

It is related by some of the old citizens, that about forty years since an old Frenchman made his appearance in Franklin, and put up at the public house of Mr. Reno. He declared that he had been a French soldier at the time of the occupation of Fort Venango, and had in fact been one of its garrison. He seemed to be an old man, and on that score there was nothing improbable in his story; moreover there was an air of honesty in his appearance and conversation that challenged the confidence of the listener. He stated that Fort Venango was hastily evacuated, and that being unable to remove all their valuables, from want of transportation, they threw them into a well that was in the middle of the fort. Among these valuables were some cannon, that were considered of much importance. Over all stones were thrown to conceal the deposits from the knowledge and possession of the Indians. The man proposed accompanying some of the citizens to examine the site of the lost treasure on the following day, but during the night he was taken ill, and, lingering some time, died without any investigation having been made. A successful search has never been made since that time.

As a relic of the past, the history of a small cannon that was exhumed about thirty years ago, may not be without its interest. It had most probably belonged to what is known as the " English Fort," as it was discovered near its side. It was found by some boys, who were making a slight excavation in the bank, by whom it was quickly brought from its long sleep to the light of day. On examination, it proved to be an antique pattern of a four-pounder, without inscription or other mark that would have proved its nationality. It had however been spiked, and the trunions broken from it, in order to render it worthless before being buried up to enter upon its sleep of a century. The old veteran, however, soon fell into the hands of a practical gentleman, who had the spike removed, and a band put around it containing new trunions and then mounted on a carriage ready for new duty. The after history of this gun was a tragic one. It was brought out for the purpose of celebrating the anniversary of the Nation's birth, and the patriotism of the gunners being brimful, it was charged to the very muzzle with sand-stone, and, as a consequence, blown to fragments.

## CHAPTER II.

### MODERN HISTORY OF VENANGO COUNTY.

We come down to the history of the county in its relation to the United States Government. In the spring of 1787, a company of United States soldiers arrived at what is now Franklin, from Fort Pitt, for the purpose of erecting a fort to protect the country in its early settlement. The site selected for this work was a somewhat novel one. Instead of choosing a point near the ruins of the old French fort, that would have commanded both the Allegheny and French creek, they planted themselves upon the south bank of French creek about a quarter of a mile above its mouth, and just above the old French creek bridge. This work was in the form of a parallelogram, and including the outworks, enclosed a surface of about one hundred feet square. It was surrounded by high embankments of earth, outside of which a line of pickets of pine logs, was planted. These pickets were sixteen feet high. In this garrison, a force of about one hundred men was kept, until 1796, when a new fort was erected at the mouth of the creek. It was simply a strong wooden building a story and a half high, and perhaps thirty by thirty-six feet in size. It was surrounded by a picket of pine logs, but had no arrangement for the use of cannon. There was little fear now from the Indians, for the treaty of General Wayne with these tribes had laid the foundation of a lasting peace. The fort was garrisoned by soldiers, however, until 1803, when they were withdrawn altogether. These forts were usually known, the first as Fort Franklin, and the latter as "the old garrison." They have fallen to decay now. Something of the dim tracery of the former may be seen, but the latter has vanished altogether, and its site is washed by the restless tide of French creek.

In 1795, an act was passed by the Legislature to lay out a town at the mouth of French creek. This act was carried out by commissioners appointed for that purpose, General William Irvine and Andrew Ellicott, with what taste and judgment can be best determined by a reference to the map of the town. The name Franklin, was probably suggested by the name previously given to the fort. The town extends along the southern bank of French creek and western bank of the Allegheny. The valley containing the town plot is probably half a mile wide and two miles in length, surrounded on every side by bold hills rising to the height of six hundred feet above the river. It is about sixty miles from the southern shore of Lake Erie, and about seventy from Pittsburgh, by land, and about double this distance by the course of the river.

The surface of Venango county is rather broken and rough. The Allegheny river flows nearly through the centre of the county, but such is the structure of the land, that it runs towards every point of the compass in its course through the county. Bituminous coal is found in the southern and western portions, and in almost every portion the hills and high lands abound in iron ore, that was esteemed of good quality, until the richer developments of Missouri and Lake Superior threw it into the shade. Good water is found in every part of the county. Beautiful springs gush from the hill sides, and even arise on the tops of the highest hills. There is much good farming land, yet the general impression made upon the mind of a stranger, ignorant of the wealth beneath the rock, would be that Venango would not be a desirable country to settle in; and the impression in regard to Franklin a half dozen years ago would have been, that its streets would not require paving with stone, as nature was fast performing that work with a green carpet of grass.

Franklin is elevated about seven hundred and fifty feet above Pittsburgh, so that there is a fall in the Allegheny averaging about five and a third feet per mile. From Franklin to Meadville, about thirty miles by the course of French creek, there is an ascent of about one hundred and thirty feet, or four and a third feet on an average per mile. It is perhaps fifty miles from the great table land, or dividing ridge, that separates the waters of the Allegheny valley from those of Lake Erie.

Venango county was taken from Allegheny and Lycoming counties, by an act of Assembly, passed March 12th, 1800, and was organized for judicial purposes, by act of April 1st, 1805. In 1839 its proportions were somewhat curtailed by the organization of Clarion county from a portion of its eastern territory. The county now forms a very irregular figure, with many angles, and contains about eight hundred and fifty square miles. The population was, in 1800, 1,130; in 1810, 3,060; in 1820, 4,915; in 1830, 9,470; in 1840, 17,900; in 1850, 18,310; 1860, 25,044.

Besides Allegheny river and French creek, there are numerous other small streams in the county that have recently become famous, and that are destined to become historical: amongst these the most important, are Oil Creek, entering the Allegheny seven miles above Franklin. Its tributaries are Cherry Run, Cherry Tree Run, and Cornplanter's Run. Pit-hole Creek, sixteen miles above Franklin; Hemlock, twenty-one miles; Horse Creek, eleven miles above Franklin; Tionesta, thirty miles above Franklin. Still above these are West, East, and Little Hickory. The Two Mile Run, is two miles above Franklin. Below Franklin, and flowing into the Allegheny, we have East Sandy, Big Sandy, and Scrubgrass. As tributaries of French Creek, we have Patchell's Run, Sugar Creek, Mill Creek, and Deer Creek. These names, although many of them belong to insignificant streams, are yet most of them familiar terms, from Maine to the great West, and even in foreign countries.

Oil City, now a rival of Franklin, the county seat, like Jonah's gourd, has sprung up almost in a night. It is situated at the mouth of Oil creek, and lies on both sides of the creek, extending up the steep bluff, that rises on the northern side of the creek. It contains at the present time probably 3000 inhabitants, and is a place of wonderful activity and enterprise. There "oil men" most do congregate. It already contains two banks of issue and one of exchange and deposit. On the site of Oil City, a town was laid out about a quarter of a century ago, but it did not flourish, and soon fell into dilapidation and decay. It had anciently a mill, a furnace, and store, besides a warehouse and steamboat landing: but after contending for a time against unfavorable circumstances, it yielded to manifest destiny, and went quietly to sleep, until the opening of the oil business awaked it to new life and energy. On the opposite side of the river is situated Venango City, a town of much promise, and beauty of location.

Further up Oil creek, we have Rouseville, McClintockville, Petroleum Centre, and other little villages, scattered along the valley, making almost a continuous city, from the mouth of Oil creek to Titusville, a distance of fifteen miles. There are other towns connected with the oil business, as Tionesta, at the mouth of Tionesta creek, thirty miles above Franklin; Utica, nine miles above Franklin on French creek; Coopertown, the same distance from Franklin, on Sugar creek, and Waterloo, four miles from Franklin, on Big Sandy creek.

The first settlers of Venango county were a body of pioneers, and came from various parts of the country. Some were from New England, some from Wyoming valley, and many from the middle counties of Pennsylvania. They came, as a general thing, not for the sake of adventure, but with a view of becoming permanent settlers; of planting themselves upon the soil, and making it their home. Amongst the first settlers were George Powers, Edward Hale, Wm. Connoly, Col. McDonell, and Samuel Hays. George Powers came in 1787 to assist in the erection of the fort, and six years afterwards with a view of settling, and opened a store for the purpose of trading with the Indians. He was a man of great enterprise, and spoke the language of the Senecas with the ease of a native. He was on particularly good terms with Cornplanter, the famous Seneca chief. An anecdote is related of him, in connection with his traffic with the Indians. One day a particularly fine fox skin was brought in and purchased, and, as usual, thrown up into the loft of the store. A few hours later, another fine skin was purchased, soon after another, and another; until wondering that so many fine skins should all appear in one day, an examination was made, when it was discovered that but one skin had constituted the entire stock in trade, and had been quietly removed through an upper window at the back of the store, and resold again and again, until quite a revenue had been derived from it. The descendants of Mr. Powers are among the citizens of Franklin at the present time.

William Conolly came to this county in 1800, and is still living in Franklin, at an advanced age.

Col. McDowell came to Franklin sometime between 1790 and 1795, from the eastern part of Pennsylvania. His widow still resides near the place where they first pitched their tent, and is at present the oldest inhabitant of the place. Edward Hale settled here in 1797.

General Samuel Hays came out in 1803. He was born in Ireland, but came to America when eight or nine years of age. He has been an important man in the political history of the county, and has held at one time or another, almost every office in the gift of the people. General Hays was the father of General Alexander Hays, who fell at the head of his brigade during the terrible battle of the Wilderness, on the 5th of May, 1864.

The early settlement of this region was attended with many hardships and privations. The roads were in a primitive state, just as they had been left by the military authorities. Mills must be erected; iron and salt brought from a great distance, and often on the backs of pack-horses; grain was sometimes carried on men's backs to other counties to be ground, and back again; cloth must be woven and leather tanned, by domestic process alone; and all this required labor, toil, and self-denial.

Said one old pioneer, in quaint and simple language, "Me, and my woman, come out from *Pitch*burg country, with all we had on our backs, walking and driving one little cow. We wrought hard all day, and at night we had plenty of nice dry leaves, gathered up in the woods to sleep on."

The first court was held in 1805. The court-room was a log house, fronting the Diamond, and long afterwards occupied as a drug-store. It was demolished in 1863.

At the organization of the county for judicial purposes, "the old garrison" at the mouth of French creek was used for a jail. This was the only building used for that purpose until 1819, when a small stone building, with yard attached, was erected on the Diamond. This continued to be the receptacle of prisoners until 1853, when the present jail was erected, and the old building demolished. This old building was located a little to the northeast of the present court-house, the well that is still used for water being within the enclosure of the yard.

A very important feature in the modern history of this county was its iron operations. It was once thought to be particularly rich in this metal, and its business was the honor and pride of its inhabitants. If the mines were not so extensive, nor its ore so rich as more recent developments in other States, they were satisfactory in that day, and answered a most admirable purpose.

What gave a special stimulus and encouragement to the iron business was the favorable features of the tariff of 1842, preventing, or at least lessening, the importation of iron, and enhancing the price of the domestic article. Capitalists from abroad, as well as at home, were induced to embark in the business, until, five years later, in 1847, there was no less than seventeen furnaces in blast, producing in the aggregate about 12,000 tons of "pig iron" per annum, valued at that time at about $380,000. The power used in these furnaces was in every case water; the face of the country being favorable to the construction of dams, and this being the cheapest power that could be adopted. The fuel used was exclusively charcoal, and generally manufactured on the grounds adjacent to the furnaces. Bituminous coal could not be used at all, as the variety found in the county is so largely impregnated with sulphur, as to be destructive to the iron. The consequence was that the timber was mercilessly destroyed in the region near these iron works, the result of which can be seen to the present day, in the small growth of scrubby timber upon the hill sides.

Some of these furnaces used "bog ore" exclusively; yielding less metal, but the product was a very superior iron for foundry purposes. The amount of capital employed in the business, during its palmy days, is uncertain, as many operators used borrowed capital, or carried on their business through the agency of stores, the merchandize of which was bought on credit, and paid for when the iron was taken to market and sold. "A furnace store," in those days, was a peculiar institution. It contained everything that was wanted in the county, from a needle to an anchor, and at prices and in quality that would have put a modern shoddy contractor to the blush. It was a strange business, a little hard on the conscience, a little demoralizing to the ancient canons of upright dealing, and carried on with very doubtful success. It was said in those days, that the only way in which an ironmaster's probable wealth could be estimated, was by the same process by which a rattlesnake's length could be measured, that lay coiled in the bush—by waiting until he was dead and his condition brought to light. This, however, was no doubt a calumny. Many of them were noble, honorable men, and did much for the development of the county. Many men who were trying to settle farms, were at that day dependent on the furnaces for a semi-livelihood in the interim of seasons, when they could be spared from their farms. The amount of capital necessary to carry on these furnaces, was probably about $20,000 each; or at the time when most numerous, about $340,000 in the aggregate.

But a change came. In 1848 a new tariff on iron, less favorable in its protection to home manufacturers, was passed, and the business declined rapidly. Some of these furnaces lingered long in their decline. Perhaps Horse creek, on the Allegheny, struggled the longest, but it died at last, and with it the iron business in Venango county. It will never be resuscitated. But we should cherish its memory, for it served to keep us alive until the development of better things.

Another enterprise, connected with the industrial enterprises of the county, was the Franklin Rolling Mill, embracing a nail factory. This was erected in 1843, and continued in operation until 1850, when the general failure of the furnace business caused its proprietors to cease all active operations. This establishment was carried on under the firm of "Nock, Dangerfield & Co.," and was prosecuted with a good degree of vigor, accompanied by the inevitable store, which seemed part and parcel of all iron operations. Its capacity was about eight tons of manufactured iron per day, or rather five tons of bar iron and three of nails. Iron of a good quality was produced, and nails of all the various grades in use. The capital invested was about $60,000. The site is now occupied by a barrel and tag factory, that is doing a flourishing business.

Another item, illustrating the early struggles of the county, as it awaited the opening of the great petroleum business, was the history of the Franklin Canal Company. It was at first known as the French Creek Division of the Pennsylvania Canal, and was intended as a link in the chain of water communication from Erie to Pittsburgh, and thence to Philadelphia. This link consisted in building dams on French creek, and thus opening slackwater navigation between this point and Meadville, connecting with the Allegheny to Pittsburg. It was built in 1832-3. But about this time other counsels prevailed in the State legislation, and the work lost its importance in official view, so much that the work was abandoned almost as soon as it was completed. Only two boats were ever launched on its waters, one of which made two trips to Meadville, the other but one. French creek is a large and rapid stream, and subject to tremendous freshets. It was soon found that the dams were not sufficient to control it, and before long it began to fall into decay. The dams gave way, the pools were drained, and the locks became useless. It was, however, kept in running order for descending navigation for about ten years, when it was finally abandoned.

The construction of this work cost the State about one million and a quarter dollars, with about fifty thousand dollars for repairs subsequently. The tolls of course never paid the expense of receiving them. A flour mill now uses the power generated by the outlet dam; the stones that formed the locks have many of them found their way into cellar walls, and into the public buildings of Franklin, while some of them yet linger in the dilapidated works where they were originally placed.

But to Franklin and Venango county it was the fading away of a great hope, the tumbling of a grand castle, that once promised so fairly and beautifully in the way of future prosperity.

Franklin was formerly considered the head of steamboat navigation on the Allegheny. At one time the navigation was the source of great convenience and profit. But it had greatly fallen off, at a period just before the opening of the oil trade. The iron business had ceased, and navigation was uncertain, owing to the operation of local causes on the river itself. The forests at the head waters were falling before the axe of the lumberman and farmer; the swamps were drying up, and the volume of water materially lessened. The first steamboat that was seen in Venango county made its advent in 1826. It was called the "Duncan," and was hailed with almost as much enthusiasm and gladness, as was the first boat propelled by steam on the Hudson. Tradition still relates, that as it rounded the bend in the river below Franklin, and steamed proudly up to the landing, miserable little stern-wheel as she was, that the welkin rang with the shouts of the multitude. It was on a quiet Sabbath afternoon, but the occasion was so important, and the era so wonderful, that nearly the entire population turned out to do honor to it. And this dream, too, vanished. The hope of a great internal trade, through the Allegheny and the Mississippi rivers, settled down, as others had done before it, in darkness and disappointment.

And last of all, the lumber trade that nurtured and strengthened all the northern portion of the county, began to decline. The pine trees along the margins of the streams had disappeared, and the finest portions of the timber lands were deprived of their beauty and glory, and the whole trade was declining. The Venango County Railroad, that had galvanized the county into a spasmodic kind of life, lingered along, tangible only on the imposing map of "The Great Air Line Railroad," and finally gave up the ghost. Hope was just ready to die out in the hearts of the people, and it lacked but one more feather to break the camel's back. It soon came. The crash was terrible.

This was the great frost of June 5th, 1859. It inflicted a blow upon this and surrounding counties, under which they reeled and tottered, as from an earthquake. Ruin and desolation seemed staring us in the face. The breath of God, usually so pleasant and balmy at this season of the year, was changed to frost, and the fields that were beginning to smile for the harvest were blighted, and all our hopes crushed and trampled in the dust. The memory of that pleasant Sabbath morning in June will never pass away from those who witnessed it. The sweet sunshine was all around, and the calm blue sky seemed to bend over and bless the earth; but the fields were blackened and blasted, as though the Sirocco had swept through the valley, and lingered on the hill sides. Men's hearts failed them for fear—suffering and famine were just ready to stalk forth with their gaunt and shadowy train. A deep and heavy gloom settled over the entire community, and the fear arose that multitudes of our citizens would be forced to emigrate to more propitious soils and more generous climes. In fact, many of the citizens of the county were making arrangements to leave for other regions of country, when the change came, and a beneficent Providence opened up stores of wealth, such as are unknown to any county in the Union.

It seems that for more than forty long years, this region of country has been waiting, like Mr. Micawber, "looking for something to turn up." But it did seem as though it waited in vain. One star after another would come out from the darkness, and kindle in its brightness, until it appeared as though it would light up the whole heavens with its glow; but ere this was accomplished it would die out, leaving the heavens darker than ever. But man's extremity is God's opportunity. In the August succeeding this dreadful frost, the first grand development was made of our internal resources. And the cry was like the cry for bread to the famishing prisoner. It kindled up new life; it inspired new hope. The tide became one of immigration, as multitudes suddenly sought the new found oil regions.

The chart upon the following page, giving the distances from Franklin to all important points in the oil region, will be found interesting. It was prepared by Mr. Jo. H. Simonds, of the firm of Jo. H. Simonds & Co., an enterprising and reliable firm of real estate agents in Franklin.

## UP THE ALLEGHENY.

| | | Tributaries | Miles. |
|---|---|---|---|
| Franklin to | Two-mile Run | Reed Run | 2 |
| " " | Oil Creek, | Cornplanter's Run<br>Cherry Run<br>Cherrytree Run { Michael's Run }<br>Bennyhoof Run<br>Pioneer Creek<br>Bull Run<br>Pine Creek | 7 |
| " " | Horse Creek | | 10 |
| " " | Panther Run | | — |
| " " | Pithole Creek... | Little Pithole,<br>Simond's Run, | 16 |
| " " | Muskrat Run | | 18 |
| " " | Hemlock Creek | | 21 |
| " " | McCray Creek | | 25 |
| " " | Little Tionesta Creek | | 28 |
| " " | Hunter's Run | | 29½ |
| " " | Jamisson Run | | 29 |
| " " | Tionesta Creek.. | Coon Creek,<br>Ross Run,<br>Salmon Run, | 30 |
| " " | Little Hickory Creek | | 34 |
| " " | West Hickory Creek | | 37 |
| " " | East Hickory Creek | | 37 |
| " " | Maguire Run | | 40 |

## DOWN THE ALLEGHENY.

| | | Tributaries | Miles. |
|---|---|---|---|
| Franklin to | Lower Two-mile Run | | 2 |
| " " | East Sandy Creek. | Shaw's Run,<br>Hall's Run, | 5 |
| " " | Big Sandy Creek.. | Ridgeway Run,<br>Trout Run,<br>South Sandy,<br>Little Sandy, | 11 |
| " " | May's Run | | 15 |
| " " | Big Scrubgrass Creek | | 18 |
| " " | Schull's Run | | 24 |
| " " | Little Scrubgrass Creek | | 30 |

## UP FRENCH CREEK.

| | | Tributaries | Miles. |
|---|---|---|---|
| Franklin to... | Sugar Creek.....<br>Patchell's Run.. | Homan Run,<br>Worden Run,<br>Bowman Run, | 2 |
| " " | Mill Creek | | 9 |
| " " | Deer Creek | | 11 |

# CHAPTER III.

## ANCIENT HISTORY OF PETROLEUM.

WE come now to speak of the grand and important production of Venango county. It is called Petroleum from the Greek words *Petros* and *elaion*, signifying rock oil. Sometimes it is called simply rock oil: formerly it was generally known as Seneca oil, being brought to the notice of the early settlers by the Seneca Indians. It is no new article in the world's economy. It is almost as old as the history of time. It has been found in various forms, and used for various purposes; but in all has ministered greatly to the wants and comforts of mankind. We read of it in the sacred writings, the earliest records of man. Profane writers allude to it, and mention the important uses to which it has been applied. It appears to have been known anciently more in the form of asphaltum than in any other; and the first recorded use was in masonry. We have the first record of its use, four thousand years ago, in the erection of the tower of Babel. Genesis xi. 3. The projected building was to be on the plains of Shinar. It was to reach to heaven and stand as an eternal monument of the wisdom, and skill, and forethought of its builders. Again, nearly two hundred years later, we read in Genesis xiv. 10, that "the vale of Siddim was full of slime pits," referring without doubt to the springs and fountains of asphaltum, that abound to this day in that region, and on the shore of the Dead Sea.

It was also an important material in building the walls of ancient Babylon. The work was at least in part accomplished not less than 3000 years ago. Herodotus, the Greek historian, uses the following language in relation to the buildings of Babylon: "Digging a fosse or ditch, the earth which was cast up they formed into bricks and desiring large ones they burned them in furnaces, using for lime, or mortar, hot asphaltas, or bitumen." He relates further, that this bitumen was brought from the river Is, a tributary of the Euphrates.

Cartwright, an old traveler of the last century, gives the following account of his observations at this same river Is. "From the ruins of old Babylon, we came to a town called Ait; (the modern Heet) near unto which town is a valley of pitch, very marvellous to behold, and things almost incredible, wherein are many springs throwing out abundantly a kind of black substance, like unto tar, or pitch, which serveth all the countries thereabout to make staunch their barks and boats; every one of which springs makes a noise like a smith's forge, which never ceaseth night or day, and the noise is heard a mile off, swallowing up all weighty things that come upon it."

A later traveler, Mr. Rich, says, "The principal bitumen pit at Heet has two sources, and is divided by a wall in the centre, on one side of which the bitumen bubbles up, and on the other the oil of Naptha."

Curtius, Diodorous, Siculus, Bochart, and Josephus, all speak of bitumen as forming a constituent of those mighty walls, and lofty towers, and pensile gardens, that were the wonder of the world.

As even more suggestive of some of the scenes of modern times in the Venango Oil region, we quote from Layard's "Nineveh and Babylon," giving an account of a bitumen pit on fire. "Tongues of flame and jets of gas, driven from the burning pit, shot through the murky canopy. As the fire brightened, a thousand fantastic forms of light, played amid the smoke. In an hour the bitumen was exhausted for the time, the dense smoke gradually died away, and the pale light of the moon shone over the black slime pits."

And now, after the lapse of thirty-five centuries, with all that time can do in corroding, and undermining, and destroying the work of man, the remains of these petroleum-built walls and towns still exist. Occasional fragments of bricks, with the asphaltum still clinging to them, are exhumed and brought to light.

This substance was also used in Egypt at an early day,—as early as history furnishes us the facts of the times. From the accounts, at the close of the book of Genesis, of the embalming of Jacob and Joseph, we infer that the matter of embalming was a common process then, a period of 1700 years before Christ. Dr. Pettigrew, in his "History of Egyptian Mummies," tells us, that many of the mummies that he exhumed, had the cavities of the bodies filled with asphaltum. A French writer, on the same subject, quoted by Pettigrew, says these bodies were often immersed in liquified pitch, a composition formed of common pitch and asphaltum. Modern research and observation, would confirm this notion of the extensive use of petroleum in the process of embalming. The color, the odor, the inflammable nature of the mummy, all indicate its presence. The cerements, and even the embalmed body itself, often assist in kindling and keeping up the fire of the wandering Arab, and sometimes that of the more civilized traveler.

It was used in other processes in Egyptian art. An antiquarian friend* relates, that having received a piece of Egyptian papyrus, with characters inscribed thereon, he placed it in the crown of his hat as the most secure place of carrying the precious relic to his rooms. On removing the papyrus from his hat, he discovered quite an odor of petroleum, that had been set free by the heat of his head. It is probable that in the manufacture of the papyrus, petroleum had been used in part as an agglutinant, and in part to prevent the attacks of insects, and the corroding effects of time.

From those days so ancient, that history would be dim and obscure, were it not for light from the sacred page, down to the present time, petroleum has occupied a place in the arrangements of man, either as an article of utility or luxury. It has been found cropping out, in some form or other, in every continent and almost on every island, giving token of its presence, and dimly foreshadowing its future importance and value.

In more modern times it has been found in Burmah, in Zante, in the north of Italy, and, in fact, almost always in the neighborhood of volcanoes; sometimes issuing through the earth, and at others bubbling up

* L. G. Olmstead, LL. D.

through the waters of the sea. Perhaps the most remarkable natural fountain of petroleum of which we have knowledge, is on the Island of Trinidad, in the West Indies. It is known among the inhabitants of the island, as the Tar Lake. Bitumen in a hot state is continually boiling up, until it has formed a lake three or four miles in circumference. In the centre, or at the mouth, of the fountain, the oil is hot and liquid, but as it recedes in every direction, it gradually cools and thickens until on the shores it becomes solid. This petroleum lake is now in the hands of a London company, who will soon bring its exhaustless wealth into the market. Humboldt also reports the spontaneous product of petroleum in the West India islands to be large, and as it runs to waste, it covers a large surface of the sea with its unctuous tide. This report was made in 1799.

Many years ago, a newspaper correspondent reported that within an hundred miles of Houston, Texas, there was a small lake or pond, a quarter of a mile in circumference, in the centre of which issues a fountain of petroleum boiling up from the bottom, evidently from some fissure in the rocks, and affording an indefinite supply.

The field is large. The source of supply exhaustless. It has evidently been a product of earth from the beginning. It has been one of God's great gifts to his creatures, designed for their happiness; but kept locked up in his secret laboratory, and developed only in accordance with their necessities. And now in our own day, and in these ends of the earth, the great treasure house has been unlocked, the seal broken, and the supply furnished most abundantly.

It has always been a feature of the economy of Providence, that the stores of his bounty are brought to light just as they are needed. The minerals of earth have lain hid in its bosom until absolutely needed. Men have walked over the gold fields of the world, without discovering their value, until, in the scheme of Providence, some great change was about to take place, requiring the treasure, and then it has been brought to light. Men walked over these hills and through these Venango valleys ignorant of the precious treasure that flowed beneath, even though it was suggested by a thousand bubbling springs by the wayside, and a thousand rainbow tints upon the surface of the creeks and rivers. But, during this latter half of the nineteenth century, a call was to be made upon the treasures hidden beneath the surface of the soil. A terrible struggle was approaching, in which the nation's life should be at stake; republican institutions are to be on trial. Other nations are to be spectators of the strange and terrible conflict. Not only blood but treasure is to be poured out like water in the nation's cause, and in the cause of civil and religious liberty; and so God opens the store-houses he has prepared against the day of sore trial. The gold fields of California, and Nevada, and Idaho, and Arizona pour their riches into the nation's coffers; the fountains of petroleum gush forth in wondrous exuberance; private enterprise is rewarded; the revenue of the Government is wonderfully enlarged; exports to foreign lands prevent the drainage of the precious metals from our midst, and there is kept up a just equlibrium in the trade of the country. Who can doubt but that in the wise operations of God's Providence, the immense oil resources of the country have been developed at this particular time, to aid in the solution of the mighty problem of the nation's destiny? The treasures of the earth all belong to the Lord, and he uses them for his own wise purposes, and for the promotion of his own glory.

---

## CHAPTER IV.

### ANTIQUITY OF THE OIL BUSINESS IN VENANGO COUNTY, PENNSYLVANIA.

THE oil region of Western Pennsylvania, and particularly Venango county, is the portion of oil producing territory that now occupies the largest share of attention. It is true that oil wells are successfully worked to a limited extent in some other counties, still Venango county has, thus far, monopolized almost the whole number of really valuable oil producing wells in this region.

As a product of this portion of the country, this remarkable substance became known to civilization sometime about the middle of the last century. The chief source of supply was on Oil creek, and was collected in small quantities by the Seneca Indians, and used by them in mixing their war paint and in the medication of wounds. In the latter capacity it became known to the earliest white settlers, and was used by them for almost all external injuries, as well as taken occasionally as an internal remedy. Still the supply was necessarily limited, as we shall hereafter have occasion to notice.

There are some strange facts in connection with this region, that point to a history all unwritten save in some few brief sentences, in pits and excavations, of oil operations along the valley of Oil creek, and near its mouth on the Allegheny. These detached fragments, like the remains of the Sybilline oracles, but cause us to regret more earnestly the loss of the volumes that contained the whole record. A grand and wonderful history has been that of this American Continent; but it has never been written in the archives of time. The actors in its shadowy scenes have passed away in their shadowy grandeur, leaving but dim footprints here and there to tell us they have been, and cause us to wonder at the mystery that veils their record, and to muse upon the evanescent glory of man's earthly history.

Along the valley of Oil creek are clear traces of ancient operations. Over sections embracing hundreds of acres in extent, the surface of the land has, at some remote period of time, been excavated in the form of oblong pits, from four by six, to six by eight feet in size. These pits are often from four to six feet still in depth, notwithstanding the action of rain and frost for so many years. Some of these pits appear to have been of a circular or oval form, but all to have been excavated with care, and with reference to one design. They are found in the oil region, and over the oil deposits, and in no other place; affording unmistakable evidence of their design and use. The deeper and larger pits appear to have been cribbed up with timber at the sides, in order to preserve their form, and better to adapt them to the end in view; this cribbing was roughly done. The timber was deprived of its bark, halved, and rudely adjusted at the corners. In one instance, as workmen were excavating the earth preparatory to the erection of a saw mill, in a soft, marshy place, one of these circular pits was discovered in the form of a well, perhaps four feet in diameter, with the walls lined with timber set up vertically. These timbers were twelve feet in length, indicating a well of that depth. This well, of course, was filled up nearly to the surface with mud and sediment; but indicating the same design as those before described. The timber had the bark removed, but was apparently sound and free from decay. In the immediate neighborhood of this well there is about an acre covered with these ancient works. In one of these a tree was felled, upon the stump of which eighty concentric circles or growths were counted, indicating its probable age. This was half a century ago. This record of the forest trees is not unfrequent among these oil pits. Farther up the creek, upon the septa that divide them, and even in the pits themselves, trees have grown up more than one foot and a half in diameter, with as many as two hundred of these growths, indicating an antiquity ante-dating the earliest records of civilized life in this region. For centuries has this treasure been affording intimations of its presence. Before Columbus had touched these Western shores, was it gathered here, in this valley, as an article of utility or luxury, by the processes of design and labor, and with the idea of use, traffic, and emolument.

By whom were these excavations planned and these pits fashioned, that tell of the search for, and the collection of, petroleum so many years ago? Let the mighty dead that are slumbering in our valleys, and the remains of whose fortifications and cities are yet scattered all over the great West, as magnificent, as vast, and gorgeous as the ruins of Nineveh and Babylon, arise and speak, for they alone of mortals can tell. Here is a question for the antiquary quite as interesting and greatly more mysterious than any that pertains to the ruins of ancient Europe, or Asia, or Africa. Concerning the latter we have dim and misty records. The memory of the past is kept alive either by historical records, or by reasonably reliable tradition. We know who built the walls of Babylon, and sculptured the winged lions of Nineveh. We are acquainted with the people, their language, and manners, and customs, who lived in Pompeii and Herculaneum. But of those who toiled, and lived, and loved, and died in these oil valleys we know absolutely nothing. Their origin, their race, their language, their religion is shrouded in darkness and mystery that is impenetrable. We have but the dim footprints, we can measure them approximatively; but cannot tell whence they came, or whither they have gone. Not a date has survived them, not a tradition has remained among any people that have survived them, that would satisfy us as to their history.

From the fact that some of these pits have been cribbed with timber bearing marks of a cutting instrument in its adjustment, many have assigned them a modern origin, and supposed that their construction was due to the French, who, at one time, occupied to a certain extent the Venango oil region. But this theory is scarcely plausible, and certainly not tenable. Fort Venango was completed by them on the present site of Franklin in 1754, and this was probably about the beginning of their active operations in this region. But the construction of these pits no doubt ante-dates the French operations very many years. Timber placed in these oil pits, and thoroughly impregnated by its preserving properties, would be almost proof against the ravages of time. As evidence of this, petroleum, in some of its forms, entered largely into ingredients used in embalming by the ancient Egyptians; their embalmed bodies remain perfect to this day. Even the cerements remain distinct and perfect in every fold, and crease, and thread, as they were arranged in days when Israel was toiling under the oppression of the taskmaster, rendering day by day "the tale of bricks."

There is evidence, too, from the growth of timber in the very beds of these excavations, that they claim an antiquity greater far than the occupation of these valleys by the French. There are as large trees growing among and upon the oil pits, as in the surrounding neighborhood, and there is no special reason to suppose that these were the first trees that sprung up after the abandonment of the work. Those now found there were, some of them, not less than two hundred years old, and it would be as unreasonable to suppose this period marks the date of the construction of these works, as that the reputed age of the oldest trees on any of our table lands marks the date when the country emerged from choas. These trees, however, undoubtedly indicate that these operations were not carried on later than their growth would indicate. Year after year, a silent, solemn record was made by the concentric circles, first in the shrub, next in the sapling, and then in the fully developed tree, that tells of the lapse of time since these mysterious works were in operation.

Besides this, where was the market for the immense

quantity of petroleum that must have been produced from these excavations, on the supposition that they were constructed by the French? Surely not in Canada or France, for neither in the misty traditions, nor early records of that time, do we find reference to any large quantity of this product; nor had they facilities for conveying it to the seaboard, had there been a demand for it at home.

The sole object of the French at that time was to gain military possession of the country. This was the declared object, as seen from their records and military correspondence. It is seen in the line of forts thrown across the country from Erie, Pennsylvania, extending through Waterford, (Fort Le Boeuf,) Franklin, (Fort Venango,) to Logstown, a point on the Ohio river below Pittsburgh. There is no evidence that they made any attempt, either to cultivate the soil, save in a temporary way, or to develop the mineral resources of the country. Another objection that is fatal to this theory is, that, at this time, the Indians were exceedingly jealous of the encroachments of the pale faces. They watched their movements with the most unslumbering vigilance. The homes of their children, and the graves of their fathers were in danger; and it is beyond the bounds of credulity to suppose that they would have permitted them to carry on these operations for years, turning up the soil, cutting down the timber, and desecrating their hunting grounds, when their overpowering numbers could easily have prevented it at any time. At a much later day the Indians claimed this unguent as one of the special gifts of the Great Spirit to his red children, and they would have as readily tolerated the driving away of their game, and destruction of their corn, as the carrying away the "medicine" given them to heal their wounds, anoint their joints for the wilderness march, and adorn their bodies for the war path.

Another theory, that has been somewhat popular, is, that these pits are due to the labors of the American Indians. But the very term labor seems absurd when used in reference to these lords of the forest. They never employed themselves in manual labor of any kind. They scorned it as unworthy their dignity and independence. The female portion of the community planted a little corn, and constructed rude lodges to shelter themselves from the wintry blast; but they never even dreamed of trade or commerce. The Indian loved to roam through the wilderness, to hunt the red deer and follow the war-path—to seek for game to supply present wants, or to bring home the scalp of his enemy as a trophy of his prowess; but would scorn to bend his strength to rude toil in excavating multitudinous pits for the reception of oil, or in bearing it from place to place after it had been secured. They had no implements in their possession when first known to the civilized world, either for excavating or for cribbing the work when excavated, and it is preposterous to suppose that their civilization was of a higher type, or their knowledge of the arts more extensive at any former period of their history.

Beyond all doubt the Indians were well acquainted with the existence and many of the properties of petroleum. That they valued it is beyond a question. They used it both for medicinal and for toilette purposes. But they knew of its existence and production just as the earlier white settlers did: they found it bubbling up from the bed of the stream, and from the marshy places along its banks. They, no doubt, collected it in small quantities, without labor and without much forethought, and with this small supply were content. The surface oil would more than answer all their purposes. But even if a much larger supply had been desirable, and if the modern idea of traffic had found a place in their hearts, they had no facilities for conveying it from place to place. Even at the present time, with all our improvement in the arts, and with all the stimulus of enterprise and demand, the great desideratum is an appropriate vessel for carrying petroleum from place to place, or retaining it safely in any locality; but the Indians were utterly destitute of any appliances suitable for the purpose. If they were acquainted with a rude kind of pottery, it was without glazing, and unsuitable for containing fluids, particularly petroleum. Fragments of pottery are still found in these valleys, probably due to the labor of the Indian squaws, but it is brittle and exceedingly porous—hardly capable of being used as drinking vessels; and we have no knowledge of their ability to construct vessels of any other material that would answer the desired purpose. The inference is therefore fair, that, for purposes of trade, the production of oil was not desirable in so large quantities as indicated by these excavations.

The same reasons would hold good in reference to its use in the religious ceremonies of the Indians. It could be used only in limited quantities, from the want of convenient receptacles for its retention and carriage. Besides, we doubt whether the Indians were sufficiently devout to resort to such labor and pains in religious worship. Nor would the Indian character readily display itself in this form of worship. It partakes too much of the ideal and sentimental. They sometimes worshipped the sun, but the oblations were generally in the form of a sacrifice laid upon the altar. The victim usually offered by the Senecas was a white dog, and there is no reference in any well established account of anything like fire-worship, a form that seems to have been Persian in its origin.

In support of the theory of Indian origin, reference is sometimes made to a letter said to have been written by the commander of Fort Du Quesne (Pittsburgh) to Gen. Montcalm, describing a grand scene of fire worship on the banks of Oil creek, where the whole surface of the creek, being coated with oil, was set on fire, producing in the night season a wonderful conflagration. But there is room for the suspicion that this account is apochryphal. Such scenes as are there described have been witnessed on Oil creek since the beginning of the modern oil trade. During the continuance of several accidental conflagrations, the scene has been awfully grand and impressive. It has been strongly suggestive of the conflagration of the last day, when

"The lightnings, barbed, red with wrath,
Sent from the quiver of Omnipotence,
Cross and recross the fiery gloom, and burn
Into the centre;—burn without, within,
And help the native fires which God awoke,
And kindled with the fury of his wrath."

But these terrific scenes were when hundreds of barrels of oil had been stored up in tanks, and when the combustible fluid was spouting from the wells, mingled with columns of gas, at the rate of many hundred barrels per day. Before the present deep wells were bored, oil was not produced in sufficient quantities to cause such a conflagration, and previous to the date of modern oil operations, there was never seen upon the creek a stratum of oil of sufficient consistency to be inflammable.

The remains of the once powerful confederacy of Indians known as the "Six Nations" still linger in Western Pennsylvania, in a region not very remote from Oil creek, and at the time this region was settled by the present inhabitants were found in great numbers, but they can throw no light upon the origin of these pits. In regard to their history they can give no more information than they can concerning the mounds and fortifications, ruined castles and dismantled cities, that tell us of a once glorious past, of a mysterious decadence, and of the utter vanity of all merely earthly glory.

There are men still living in the oil valley who were on terms of familiar intimacy with Cornplanter, a celebrated chief of the Senecas—the last of a noble and heroic line of chieftains that had borne sway from the Canadas to the Mississippi,—and who was living at the time of the French occupation of the country. He had been in his time a mighty man of war—was born about 1735—was with the French and Indians at Braddock's defeat in 1755—at the Wyoming massacre; but afterwards became the friend of the white man. His home during the last years of his life was on the Allegheny, about seventy miles above Franklin. Here he died in the year 1840 or 1841, at the age of about one hundred and five years. Cornplanter, although allied to the white race, and having their blood in his veins, was essentially an Indian in all his feelings and instincts. He was an eloquent orator, a shrewd tactician, a bold warrior, and possessed the entire confidence of his own people. He was one of the great men of the dusky races that are now so rapidly passing away. For nearly a century he had had intercourse with the chiefs and braves of different tribes, and was well versed in their traditionary lore; but in reciting his own deeds and memories, and those of his fathers who had gone to the silent hunting grounds of the spirit land, he could say nothing of early oil operations, any further than the collection of it in small quantities for medicinal or ornamental purposes. Of French operations on Oil creek he could say nothing, and on the origin of these pits he could throw no light. This would be inexplicable on the theory that they were due either to French or Indian labor. And finally these ruins presented the same appearance precisely to the early settlers, who appeared amid these scenes seventy years ago, that they do now. They appeared as ancient then as now—the same traces of a bygone age, the rounded outline, the absence of any recent traces of labor, the luxuriant growth of timber on the pits, and on the ridges that separated them—there has been little change.

The only rational conclusion, therefore, at which we can arrive in regard to these early oil operations is, that they are due, not to the Indians, or French, or early white settlers, but to some primitive dwellers on the soil, who have long since passed away, leaving no written records to tell of their origin or their history, but stamping the impress of their existence on our mountains and in our valleys, assuring us of their power and the magnificence of their operations, yet leaving us to wonder that such strength could fail, that such magnificence could perish, and that such darkness could settle over the memory of a great people.

There are such remains of antiquity scattered all over

this continent; sometimes they are of a military character, sometimes of a peaceful nature. In all the ordinary trees of the forest are found growing upon and around them, apparently of the same kind and of the same age as those found in the unbroken forest. We can ascribe them all to the same age and the same origin—the age an indefinite one; the origin, a race of people who, for want of a better name, are sometimes called "mound-builders."

---

## CHAPTER V.

### MODERN WAY OF COLLECTING OIL.

As before intimated, petroleum was found in Venango county by the earliest white settlers, and was esteemed for its medical properties. But it was found only in minute quantities. It was found in particular localities along Oil creek, in the town of Franklin, and other places along the banks of the Allegheny, just where there happened to be fissures in the rock permitting it to escape. It would be found issuing with the water from springs, sometimes bubbling up from the bottom of the river in small globules, that, rising to the surface, disperse themselves upon the water and glide away in silent beauty. Many such indications were seen in the creeks and rivers. A drop of oil would escape from the rocks or gravel beneath, and, accompanied by its gaseous attendant, would appear like an air-bubble until it reached the top, when a beautiful surface would be presented on the water, reflecting all the colors of the rainbow. These hints had been indicating the presence of the immense treasure house beneath for ages, yet no wild dreamer seized the idea, for the time had not yet arrived for the revelation and the use.

Many families drew their supply from little springs of water on the river bank, contiguous to their dwellings. This was done by simply damming up the little spring and skimming the oil from the surface when a thin stratum had accumulated. But the principal oil spring, or that from which the largest quantity of petroleum was collected, was located on Oil creek, about two miles from its mouth. This was on the "McClintock Farm," so much celebrated in the history of modern petroleum operations, and whose possession at the present time would be equal to the patrimony of the great "king-maker," the last of the Barons. From this point the main supply was drawn for the wants of the earlier inhabitants. And as the demand was limited, no great amount of enterprise was called forth in its production. The capital invested was small, the labor demanded in carrying it on was limited in its amount, the *modus operandi* was most primitive, and yet, withal, the results were satisfactory.

A point was selected where the oil appeared to bubble up most freely, when a pit was excavated to the depth of two or three feet. Sometimes this pit was rudely walled up, sometimes not. Sometimes it was near the edge of the water on the bank of the stream, sometimes in the bed of the stream itself, advantage being taken of a time of low water. In these pits the oil and water would collect together, until a stratum of the former would form upon the surface of the latter, when a coarse blanket or piece of flannel was thrown in. This blanket soon became saturated with oil, but rejected the water. The blanket was then taken out, wrung into a tub or barrel, and the operation repeated.

But the people of that day knew nothing of oil excitements. There was little consumption. The demand was limited. It was always a drug, in the figurative sense, in the market. Most families through the country kept a supply for their own use; yet for all ordinary purposes, a pint bottle was sufficient for a year's consumption. Every good housewife was supposed to have a small store of "Seneca Oil," as it was popularly called, laid by in case of accident, for the medication of cuts, and bruises, and burns; and not even the most popular of the nostrums of the present day is so much relied on as was this—Nature's own medicine—by the earlier settlers in those valleys. It was carried abroad in small bottles to distant neighborhoods, as a sovereign remedy for

> ——"The thousand natural shocks,
> That flesh is heir to,"

until eventually it was purchased by the druggists, put up in small vials and labeled sometimes "British Oil," sometimes "American Oil," or "Rock Oil," according to the popularity of the terms at the time and place.

The first shipment of petroleum was to Pittsburgh, and on this wise. Mr. Cary, one of the first settlers on Oil creek, possessing perhaps a little more enterprise than his neighbors, would collect or purchase a cargo of oil and proceed to Pittsburgh, and exchange it for commodities needed in his family. This cargo consisted of two five gallon kegs, that were slung one on each side of a horse, and thus conveyed by land a distance of seventy or eighty miles. It was a small beginning, but "tall oaks from little acorns grow." The little one has become a thousand. From the same region are shipped now about thirnteen thousand barrels per day.

Sometimes the market in Pittsburgh became very dull, for a flatboatman would occasionally introduce a barrel or two at once, that he had brought down on his raft of lumber or logs. At other times the demand fell off, so that the purchase of a barrel was hazardous. On one of these latter occasions, a Yankee having taken down a small quantity, and not finding a purchaser, employed a friend to go around the druggists in the evening and inquire for Seneca Oil, and manifest a strong anxiety to obtain it. The result was, that on calling the next morning, the Yankee found a ready sale for his stock on hand.

At a period somewhat later than this, General Samuel Hays, who settled here in 1803, relates that at one time he purchased all the oil produced in the country, and that the highest annual yield was sixteen barrels. This oil he sold at the time in Pittsburgh at about one dollar per gallon.

In the meantime a well was bored on the bank of the Allegheny, near the mouth of Horse creek, twelve miles above Franklin, in quest of salt water, with the view of manufacturing salt. This was some forty years ago. After sinking the well through the solid rock to the distance of seventy or eighty feet, oil presented itself in such quantities, mingled with salt water, as to fill the miners with disgust, and induce them to abandon the well altogether. They were boring for salt, not petroleum. Salt was an article of utility and large demand, and would be remunerative; oil was of comparatively small importance, and already a drug in the market through the spontaneous yield of nature, and so the well was abandoned as an intolerable nuisance.

Another link in the chain: about thirty years ago a well was dug in the town of Franklin for the supply of a household with water. At the depth of about thirty feet, there were evident signs of petroleum. The supply soon became so great as to be annoying to the workmen, and by the time it was completed the clothing of the principal man engaged was so saturated with oil that he went to the creek to indulge in a bath, and, at the same time, send his oily clothing down the current as worthless. A hole has since been sunk through the rock in the bottom of this well, but the success has not been as good as in other wells in the neighborhood. The depth, however, was not as great as in some others.

Another instance still is recorded on the opposite side of French creek. It was in constructing the lock for the Franklin canal. In the excavations that were made prior to commencing the masonry, there was the same trouble from the flow of petroleum. The masons were ignorant of the part this substance had played in the masonry of Babel and Babylon, and execrated it as a nuisance, and were almost ready to abandon the place in disgust. And in general, there were several wells in the neighborhood that bore a trace of oil both to the smell and taste, that was absolutely offensive to those unaccustomed to them.

In all the cases cited above, were strong hints of the existence of the treasure concealed in the rocks beneath, and even of the manner of obtaining it. The treasure was, in fact, knocking at the door, and asking to be released, in order to contribute to human enjoyment and wealth. But the time had not arrived for the enjoyment of this great boon, although it was approaching step by step. The earth was at first made the repository of all the gifts that man should need until the end of time, either in a perfect or crude form. But they were not all revealed at the first, nor to succeeding generations, until the fitting time arrived, and man's necessities, and the world's condition, induced the great Giver to unlock the treasure house, and dispense the rich bounty.

Before man was created, the great store-house in the earth's bosom was filled with its minerals, and as the centuries rolled by, in their slow and solemn march, these treasures were gradually brought to light. Not at once did the earth disclose her mighty resources; but just as man needed them, and as they should tend to his own best interests, and the glory of the great Giver. Even on the banks of the river that watered the terrestial Paradise gold was found, yet, although "the gold in that locality was good," it was brought to light in limited quantities. In the same sacred locality, and at the same early day in the history of time, "the bdellium and onyx stone" were found in their beauty, yet were they few and rare until God would consecrate the treasures of the earth to his own service in the construction and adornment of the Tabernacle and Temple. The great treasure house was then opened until gold became common as brass, and precious stones numerous almost as the pebbles of the brook, and the riches of the earth became eternally consecrated to the service of God.

In the present century, and also within our own recollection, when the world's business seemed to be stagnated; when the sails of commerce flapped idly at the mast; when the great highways of trade and traffic were in danger of becoming deserted, and the coffers of the nation were almost exhausted, and fearful times of war and peril to the nation's life impending, Providence unlocked the golden treasures in our Western States and Territories, and every department of business has become prosperous, and every branch of industry received a new impetus. A new lesson has been taught the world: that God's treasures are inexhaustible, and that his hand can never be shortened.

And now here, in this remote county of Western Pennsylvania, so humble and so poor in agricultural resources, God's treasure has been concealed for ages, locked up in the very heart of the eternal rock, awaiting the time of need, and accomplishment of the eternal purposes of Omnipotence. It has oozed forth, in limited quantities, during the lapse of centuries, as if to show us now, that man cannot lay his hand upon the house of God's treasures until his own appointed time.

We know not where the great Chemist has his laboratory, or where he formed the treasure; most probably they were fashioned when the earth assumed its present form; and since "the morning stars" sang creation's hymn together, deep down amid earth's rocky caverns, through the revolving centuries, the stores have been accumulating that are destined to bless the world, and become elements of national wealth.

And now, from that great laboratory, through innumerable channels, cut through the living rock by the Creator's hand, and by "paths which no fowl knoweth, and which the vulture's eye hath not seen," is that treasure brought to the earth's surface, just in our time of need. When other supplies are failing, and other resources giving way, we see God's wisdom manifested in opening up new channels. The great Benefactor would teach us that however straitened we may be, he is never confined, that his resources are unlimited, that for every emergency in our history there is provision made, and that our time of need is but the beginning of his overflowing bounty.

## CHAPTER VI.

### FIRST IMPORTANT DISCOVERY.

It is really strange how slow men were to discover the abundance of this supply, and to trace it to its luxuriant deposit amid the rocks. While it was literally forcing itself upon their observation, it was only by a roundabout process they discovered its richness and importance. It had been hitherto used almost exclusively as a medicinal agent, but gradually necessity was forcing it into use as an illuminator under certain circumstances. Above and around the oil valley the lumber business occupied the attention of many. In the saw mills work was often carried on at night, rendering artificial light necessary. Some practical sawyer resolved to try the experiment of burning Seneca Oil, as it was still called, in the mill. The mill being an open structure, the huge volume of smoke, that arose from the flame was no particular objection, and the light was tolerably satisfactory. Thus the pressure began to be applied to the idea, and the stimulus was having its effect.

Perhaps the first real conception of the modern petroleum business, had its origin in the mind of a young physician in the Venango region. It was certainly a natural one, yet withal original. Yet it was but a dream, and, like many another dream of the past, it was in advance of the age, and resulted in nothing but speculation. From boyhood he had been familiar with oil developments, as they have been described in these chapters, and now that a demand was increasing above the supply, his mind was attracted anew to the subject. In looking at the numerous slight veins of oil that oozed up along the bed of Oil creek and the Allegheny river, the thought occurred to him, that by tracing these little veins to their source the main artery might be reached. And as this tracing must be through the rock, the proper plan would be to bore down through it, until a large vein or perhaps the main artery was reached. Here was the whole modern oil business in a nutshell—the bud nicely rolled up in its folds that has developed into such a luxuriant flower. The process of reasoning was certainly professional, and, now that it has been tested, seems a very plain, simple idea. But it was like the theory of Columbus in regard to a new continent—entirely too bold for the times, and was rejected as pure speculation. There was in this physician's theory but one link lacking in order to have anticipated the entire scheme of oil production, as it was afterwards generally carried on. The idea did not suggest itself to him that the lands along the valley of Oil creek might be leased for oil purposes, and thus his embryo idea be carried out to his profit; he thought only of purchase. And as physicians in the oil valley at that time could not command the capital that some of them can at the present, the scheme was abandoned, and the idea was lost as far as the bold originator was concerned. The idea, however, was a brilliant one, and entitles its author to be classed among the long line of those who have dreamed without realizing the vision, and who have sown valuable seed without reaping the harvest. Nor is this the only instance in which the bold thinker and the enterprising theorist have failed of entering upon the enjoyment of the substantial benefits. Those who have reaped the golden harvest, and gathered most freely the luxuriant sheaves now so thickly strewn throughout the oil valley, have generally been those who have entered second-hand upon the work, and taken the places of earlier but less fortunate laborers. Many an old well, whose derrick is now falling to decay, and underneath which lies buried, worthlessly to him, all the available means of many an honest, industrious farmer or mechanic throughout the land, will yet yield a golden harvest, when some person of capital shall have undertaken its thorough development. Many an honest, enterprising man, in traveling over the oil region, may say with the old Bard of Mantua, "Sic vos non vobis nidificatis aves."

As early as the year 1835, the presence of petroleum amid the rocks was made known on the Allegheny river, a short distance above Pittsburgh, by its interference with the salt wells, but no dream of its future importance seems to have forced itself upon either the miner or the capitalist until within the last few years. There is one notable instance of this fact. A salt well had been pumped for salt water for years, without the presence of petroleum. With the intention of securing a greater supply of water, a larger tube was inserted and a more powerful engine applied, when, to the astonishment of the proprietor, in a day or two, petroleum in considerable quantity made its appearance. Perhaps many an operator in the oil region at the present time may take a hint from this idea that will be of practical importance.

In the meantime, artificial oil had begun to be produced in large quantities from different minerals, principally, however, from cannel coal, by the process of destructive distillation. This oil was refined and deodorized, and found to be a valuable illuminator. Cannel coal is found largely in Beaver county, in the western part of Pennsylvania; also in Butler and Venango counties. It is peculiarly rich in oil, a single ton yielding by distillation forty gallons of oil. The process was carried on by placing the coal in huge iron retorts, inclined at a small angle to the horizon, and applying heat. About the year 1858, this business began to assume a vast and growing importance, and capital began to seek it out with great avidity. Cannel coal was inquired for, and lands on which it was found rose rapidly in price. In Venango, near to the oil valley, it came to the recollection of a farmer, that, in digging a well for water, the workmen had come upon a singular stratum of laminated black rock. Its value was unknown, and the only purpose to which it had been applied was in the manufacture of covers for the good housewife's milk crocks. The new inquiries brought to light the fact that the strange substance was cannel coal, and the land was sold at a good price.

A spirit of inquiry and investigation was excited. It was ascertained that this artificial oil, the product of distillation, was almost identical with the natural oil of the valleys—that the latter might be distilled and deodorized by the same processes as the other, and if found in sufficient quantities, be produced at less expense, and become a source of comfort and wealth to the country. The manner of collecting the natural oil was thought over and discussed. The mode adopted in bygone ages was considered by many—that of excavating numerous oil pits and collecting with blankets; but the process seemed tedious and expensive, and hardly remunerative; in addition to this, the finer portion of the oil was in danger of passing away by evaporation, owing to so large an extent of surface being exposed to the action of the atmosphere. As the earlier inhabitants had collected it this did not so much matter, the heavier portions of the oil being most desirable for their purposes; but as an illuminator, it was of the highest importance that the finer portions should be preserved as far as possible.

The grand idea, however, was struggling towards the light. It could not be retarded. If oil, now so greatly desired, bubbled up through concealed clefts in the rocks, why might it not be found in large quantities by boring, in favorable localities, deep into the very rock that was conjectured to be its home? Why not storm the rock-bound castle, and at once seize upon the prize? And if discovered in some localities while boring for salt water, where there had been no outside manifestations, why not expect to find it more certainly in localities where such decided surface indications had always been found?

This brings the matter down to 1853, when a new feature was developed in the business, which gradually ripened into the present system of operation. It had its origin with George H. Bissell, Esq., a gentleman of great intelligence and worth, and a graduate of Dartmouth College, N. H. From Mr. Bissell's interest and enterprise in the matter, he is justly considered to be the pioneer in the petroleum business. The inception of his interest seems to have been in this wise: Being shown a small vial of crude rock oil from Oil creek, gathered on the lands of Dr. Brewer, then of Titusville, he became greatly interested in the matter, and learning all he could of its locality and appearance, sent a young man to Oil creek to make an investigation. The report being favorable, Mr. Bissell determined to examine the subject more fully.

About this time Mr. Eveleth joined Mr. Bissell in the enterprise, when they proceeded together to Oil creek. This was in 1854, and while there they purchased from "Brewer, Watson & Co.," for the sum of five thousand dollars, the territory where the principal oil springs were found, and also procured from the same company a lease of all their remaining lands in Venango county for ninety-nine years, without royalty.

Operations were then commenced by digging pits and ditches, and pumping the oil and water into vats. This was done by water-power derived from a saw-mill.

Soon after this a few barrels of petroleum were forwarded to Prof. Silliman, of Yale College, with directions to make a thorough analysis and examination of the oil, with the view of ascertaining, as far as possible, its commercial uses and value; furnishing him at the same time with all the necessary aids, particularly a photometer on an improved plan. After some months of labor and experiment, Prof. Silliman presented, as the result of his labors, a report, which was published by Eveleth & Bissell in 1855.

They then organized a joint stock company in New York, under the style of the "Pennsylvania Rock Oil Company," but soon after this company was re-organized at New Haven under the same name, of which Prof. Silliman was elected President. Of this company, Eveleth & Bissell, from first to last, continued to hold the controlling interest. Under its auspices the plan of ditching and pumping was continued until 1857, when various plans were submitted for further development, and the question of boring an artesian well generally advocated. The directors, however, were not harmonious. Finally some members of the Board submitted a written proposition to sink a well at their own expense, provided they could have a lease of forty-five years, the lessees paying the parent company twelve cents per gallon for all oil produced. This was agreed to, and the lessees selected Col. E. L. Drake, who had purchased a small interest in the stock, as their superintendent and agent.

Soon after Col. Drake, furnished with all necessary funds, started for Titusville, two miles below which the springs were located. Operations were slow. There was neither experience nor the success of those who had succeeded before to assist. The Colonel should be recommended to the proper authorities for a Brigadier, if not a Major-Generalship, for the nomination would surely be confirmed by a congress of oil men. This boring commenced near the upper oil springs, in the northern portion of Venango county. There was a great hope here, but it was a hope that required strong faith to keep it alive. Still the Superintendent hoped on, and the partners, Eveleth & Bissell, like earnest men that they were, continued undismayed by the difficulties that gathered around. It required strong courage to begin and continue the work. The very announcement of an intention of boring into the earth, or rather into the solid rock, was sufficient to provoke mirth and ridicule. The enterprise appeared to many quite as visionary as that of Noah did to the antediluvians, in building his ark against an anticipated inundation. It was supposed by many that the object of search was salt water for the manufacture of salt. Occasionally a salt lick was found in the low lands, and this object appeared a natural one.

However others might have thought, the company and their agent thought only of finding Rock Oil, as it was generally called, with perhaps a shadow of a thought that if they failed in this salt might be found.

It was like the inception and preparatory operation connected with all great and important works here—carried on amid struggles, and discouragement, and fears. There was the grand idea struggling to the light; there was the determined purpose developing itself; there was the new era just on the threshold, and the whole rested with one earnest, resolute heart. There was the pressure of the one great thought, that by some mysterious power had got possession of the man's heart, urging him forward, and likewise on the other hand there was the dread of failure, the fear of ridicule—the shadow that falls so heavily upon men's hearts from time to time, when in pursuit of some uncertain object—and these things kept up the struggle with greater or less power. It reminds us of grand old Martin Luther at the Diet of Worms, crying out, "Here I stand. I cannot do otherwise. God help me. Amen!" So all grand ideas are wrought out, and all grand purposes carried forward under the mysterious guidance of a power, that seems invincible in its operations.

But the work went forward through good report and through evil report, particularly the latter, in spite of all difficulties, and in face of all obstacles. Slowly, patiently, hopefully, did the steel chisel work its way through the hard rock, stroke by stroke, until August 28th, 1859, when, at the depth of seventy feet, the drill suddenly sank into a cavity in the rock, and immediately there was evidence of the presence of oil in large quantities. The shout of the weary laborer, was like the cry of "Land ho!" among the weary disheartened mariners that accompanied Columbus to the Western World. And with more reason might that quiet, thinking colonel, at that moment, have rushed through the neighborhood, crying "Eureka! Eureka!" than had the old Grecian Archimedes, at the solution of his mathematical problem, as he lay amid his quiet lucubrations in the baths. The goal had been reached at last. A pathway had been opened through the rocks, leading, not to universal empire, but to realms of wealth hitherto unknown. Providence had literally forced on man's attention that which should fill many dwellings with light, and many hearts with gladness.

Upon withdrawing the drill from the well, the oil and water rose nearly to the surface. The question was now to be tested whether the petroleum would present itself in sufficient quantities to justify farther proceedings, or whether this was like many another dream, to vanish in darkness, or dissolve in tears. Perhaps, after all the labor, and anxiety, and hope deferred, the quest was like that of the treasure supposed to be hidden in the earth by pirates' hands, coming in sight only to vanish when within grasp. But the time of waiting was not long. The well was tubed, and, by the aid of a common hand-pump, yielded ten barrels per day. The dream was realized; the vision was fulfilled. Faith was rewarded. By means of a more powerful pump, worked by a small engine, the quantity was increased to forty barrels per day. The engine worked day and night, the supply was uninterrupted, and the question was considered settled.

This oil well at once became the centre of attraction. It was visited by hundreds and thousands, all eager to see for themselves, and test by actual experiment the wondrous stories that had been related concerning its enormous yield, by counting, watch in hand, the seconds that elapsed during the yield of a single gallon. Oil at this time was worth one dollar per gallon, so that the fortune of this pioneer in the business was considered settled, and impulsive men began to dream of an interest in this new source of wealth, and of purchasing shares in Colonel Drake's oil well, and from this the idea extended to the enterprise of boring other wells. The arithmetic employed was on this view; forty barrels per day were multiplied by three hundred working days in the year, making twelve thousand barrels per annum, and this multiplied by forty dollars per barrel footed up the handsome sum of four hundred and eighty thousand dollars as the annual yield of the well.

It might be added here, that the arithmetic and the facts did not agree, as the experience of oil men since will assert; but the matter looked plausible enough then. Nor was it strange that men's pulses quickened, and their hearts beat more rapidly than was their wont, at the apparent development of this new order of things.

The work now commenced in earnest. A tide of speculators and operators began to set in toward the oil regions, which would have overpowered that of California or Australia in their palmiest days. Nor did the excitement stop at the valley of Oil creek. It extended down the Allegheny to Franklin, and up to Tideoute. It was soon felt up French creek, and the Two Mile run, and eventually down the Allegheny for several miles below Franklin. Wells were sunk in all these localities, many of them yielding from two to forty barrels per day. Still the excitement was limited in its extent. It was long before it could make its influence felt in the Eastern cities. Capitalists there were slow to believe the marvellous stories that were told of the Venango oil regions, and kept aloof from the excitement. But it gradually worked its way eastward until Philadelphia, New York, and Boston, with the regions represented by them, were enlisted in the work.

It need hardly be said here that on the successful issue of the attempt to find oil in the rock, the partners, Eveleth & Bissell, at once set out for the Oil creek region. For them the vision was realized, and fortune expanded in a long, brilliant vista before them. These gentlemen continued to identify themselves with the petroleum business together, until the death of Mr. Eveleth in 1863, since which time Mr. Bissell has continued his connection with it, either alone or in company with other parties. It is a pleasant task, too, to chronicle the success of these gentlemen, not only in developing the business, but in reaping substantial fruits from their connection with it.

The report of Professor Silliman referred to, is full and clear, yet, as he had nothing but the thick surface oil before him for examination, the result was somewhat different from what it would have been, had he been in possession of the light oil now found amid the rocks. Still he recommended it as a most promising illuminator as well as lubricator.

## CHAPTER VII.

### MODE OF PROCEDURE.

THE age of Petroleum had fairly dawned. The fortune of the valley of Oil creek was now settled, and the price of lands throughout its whole extent, from the new well to the Allegheny, immediately rose to a very high figure. Sometimes entire farms were sold, but generally they were leased in very small lots. The terms of lease were at first easy, the operators giving one-fourth or one-fifth of the oil as a royalty to the owner of the soil. Gradually the terms became more exacting, until not unfrequently one-half and even five-eighths of the oil was demanded, with the addition of a considerable sum of money as a bonus. Sometimes the proprietor of the soil required the proposed operator to furnish him his share in barrels; that is, not only turning him over a third or a half of the oil, but furnishing him the barrels to contain it. With this arrangement, it afterwards come about that as the price of oil fell and the price of barrels advanced, the entire proceeds of some wells would hardly purchase barrels to contain the royalty share pertaining to the owner of the land. The consequence was, that all such wells were either closed, or the leases modified so as to be adapted to the new order of things.

These leases were usually drawn for an indefinite time, but required the lessee to commence operations by a certain time, and to prosecute the same with a reasonable amount of diligence, either to success or abandonment; and generally there was a clause to the effect that if petroleum was not found in paying quantities within one year from date of covenant, then the lease was forfeited, and the full control of the land reverted to the owner.

This matter of leasing land for oil purposes, at one time, amounted to a monopoly in some sections of the oil valley. The landholders in many places were men in very moderate circumstances. By great frugality, they had been able to live comfortably, but had no extra means with which to embark in speculations. Sometimes they had neither taste nor energy for this business, or lacked faith in the general result, but were willing that others should embark in the business by sharing the profits with them. There was no risk to the landholders, and the profit might be considerable. In this state of affairs shrewd and enterprising individuals made a business, for a time, of leasing all the lands in certain localities, with no intention of operating themselves, but with the design of sub-leasing to real operators. Sometimes these lands were leased in bulk for one-eighth or even a tenth as the share of the proprietor of the land, without any bonus; and afterwards sub-leased in small quantities, receiving one-third or one-half the product with a handsome bonus in addition.

In the earlier stages of the oil business, operations were usually carried on by large companies. By many it was considered much as an experiment, and many were unwilling to invest, save in small amounts. Others, who had but little to invest, thought they saw in these companies a mode of securing large profits in return for their small investments. The manner of forming and carrying on these companies was simple and primitive. It was, perhaps, really a partnership for the transaction of business, but was hardly so considered by many at the time. A few brief articles were drawn up, officers appointed, and a lease secured. Sometimes a member or two of the company were appointed to carry on the work. Sometimes it was given out by contract. Assessments were laid upon the members just as funds were required to carry on the work, and generally with the understanding that when these assessments were not paid within a reasonable time, the standing of the member so neglecting or refusing should be forfeited, together with any previous assessments he might have paid.

The first company organized at Franklin consisted of fifty members, the second of forty, the third of fifty-one; and the general assessments were ten dollars upon each member at a time. In those early days a well could be sunk as deep as was supposed necessary to settle the question of success or failure, for about one thousand dollars, the implements for boring and pumping costing one thousand dollars more; so that the risk was not very great, even if the prospect of a large fortune was not particularly brilliant.

These companies, at the first stages of proceedings, were not chartered, nor had they any authority of law for holding or transferring real estate, any farther than as simple partnerships for the transaction of business. Each individual member of a company was, of course, responsible for the debts of the whole concern, should the creditor demand it at his hands. And as the hunter always aims at the finest deer in the herd, if he has but a single shot, creditors usually made use of the same discrimination in serving their processes. As many of these companies, after a single partial experiment, disbanded with debts of various kinds, the matter often became exceedingly embarrassing to certain members of the disbanded association who lived near the site. Many an individual who had never in his life been visited by an officer, had that privilege accorded him, together with that of paying the debts of the defunct association.

Such companies are now usually organized under the laws of Pennsylvania (Purdon's Digest), receiving a charter, conferring upon them certain rights and privileges, and restricting them to certain courses of procedure.

These large companies, in one respect, did a good work, and in others were damaging to the general interests of the business. As to the first, they assisted in developing territory that was doubtful, and getting up an interest in neighborhoods that had not been even partially explored, thus attracting the attention of capitalists from other places, who had the means and the energy to explore fully and satisfactorily. Their operations were like the firing of the skirmish line, that reveals the location and attracts attention to the coming battle.

Yet, after all, it is doubtful whether these large companies did not do more harm than good, as far as the general result was concerned. They were most generally unsuccessful. There were too many proprietors. Too many pilots were on board the ship, and among them all, each one of whom was part owner, the ship was almost sure to be driven on shore or wrecked amid the breakers. One and another would decline paying further assessments, until a feeling of discouragement would settle down upon the company and paralyze their efforts.

Again, among so many there was no one having a sufficiently strong interest to urge the matter along vigorously and hopefully, and, in addition to all, the matter was carried on with great extravagance. A small outlay would be but little to each one of fifty members, but at the last the aggregate was sufficient to preclude the idea of dividends, even when successful in developing petroleum.

The result was, that all over the territory operated by these large companies, oil derricks were to be seen falling to decay, fragments of masonry testifying that engines had once been used there, but now removed, the ruins of oil tanks and other fixtures pertaining to the business of collecting petroleum, rendering the scene desolate and in the highest degree discouraging, save to the shrewd and experienced operator. To most persons it was uninviting, and to many repulsive, for those ruins, or "dry wells" as they were termed, were like so many stranded, dismantled ships on an iron-bound coast; they warned all voyagers to be on the alert, and avoid the fate of those who had preceded them.

After this, there was another feature attending early oil operations that was equally disastrous, at least to those engaged. Entertaining the idea already alluded to, that a certain sum, and that not a large one, would be sufficient to develop a well, many persons of small means undertook the work, sometimes singly, sometimes in connection with one or two neighbors. They calculated only on a single experiment, and their whole available means were staked upon a single well, and this perhaps not more than three hundred feet deep. The work was commenced hopefully, and carried on heroically for weeks and even months, as wells were not bored so rapidly then as now. Funds would begin to get low, courage would begin to fail, and still no oil beyond mere surface appearances. Here was a quandary. All the available resources of the man or the little company were down in the crevices of the rocks, with nothing in return. Sometimes in this state of affairs the work was abandoned, the lease forfeited, and the loss sustained with the best philosophy that could be brought to bear in the case. Sometimes the "claim" was sold to parties possessing more capital, and blessed with greater resources, and in their hands became a decided success. Instances are known where wells have been laboriously put down until means and faith both becoming exhausted, the enterprise was sold for a trifle, and in the new hands soon produced a tremendous fortune.

There was another reason for some of these serious losses to the early operators. At one period the pumping wells were all comparatively of small capacity, from two or three up to thirty or forty barrels per day. In many sections not more than eight or ten barrels were expected under the most favorable circumstances. This yield, when petroleum sold for half a dollar per gallon, would be remunerative and encouraging. But the new feature of flowing wells to be described in a succeeding chapter, changed the whole face of the oil trade, and, for a time, deranged all the plans that had hitherto worked so well. A mighty deluge of petroleum was at once thrown upon the market, bringing down the price to a merely nominal figure. A panic seized the smaller institutions, and operations were suddenly arrested. Operators who had hoped to open a well yielding from three to ten barrels per day, thought they saw the futility of farther operations, suspended work, removed their machinery, and abandoned their leases to the owner of the soil. Many an eager operator returned to his home a sadder if not a wiser man, after such experience as this, having expended possibly the surplus of his earnings on his farm or in his shop the year before, in addition to the actual labor on his oil well.

There are many instances, too, where losses more provoking still attended this new development in oil operations. Wells that had already proved a success under the old order of things, yielding perhaps two to six barrels per day, and profitable when owned by one person, or a small company whose aspirations were not extravagant, become suddenly worthless. With oil at five cents per gallon, pumping would be ruinous. The consequence was, that all small wells were immediately closed. Many of this description were incontinently abandoned, the chamber and engine removed, and the lease suffered forfeiture. Some of the more shrewd and calculating operators of this description, thinking that the new order of things would change, clung tenaciously to their modest little wells, and "determined to fight it out on this line, though it should require all summer." In some cases a modification or renewal of the lease was obtained, permitting the holder to rest until the storm should be overpast; and, in others, the well was pumped occasionally for a short time, in this way preventing the forfeiture of the lease.

The storm did expend its fury after a while, and those small wells are once more "paying institutions." The patient waiting of the small operator has been rewarded by finding his well, if yielding "heavy" oil, once more bringing him a dollar per gallon. But in the great majority of instances, those small yielding wells have gone into the possession of capitalists, and assisted to swell the mighty monopoly that is spreading itself over the oil region.

Large numbers of those abandoned wells and forfeited leases, are still, no doubt, biding their time; not awaiting further operations at the hands of their original projectors, but other and more fortunate operators. They are scattered over a large extent of territory, on every stream and in almost every valley throughout the oil regions. As a general thing they are avoided by persons seeking a location for wells, from the general impression that they are failures as oil wells. But this is not necessarily, nor will it be ultimately found generally, the case, but from causes before explained. The territory has not perhaps been fully explored. The well was not deep enough perhaps to reach the oil-bearing strata. Perhaps in cases where the well was deep enough to expect success, only one experiment was tried, or there may have been a defect in the mode of tubing and pumping. Some of the greatest instances of success have been with abandoned wells; and, no doubt, this will be the case in future. There is a wide field of operations for such enterprises in many portions of the country, where decaying derricks are now but the monuments of disappointed hopes.

During the last two years there has been a very great change in the manner of conducting oil operations. Large companies organized under the original plan, have disappeared almost entirely from the field. Parties possessed of small capital are not often seen competing with those of larger resources, and, as a general thing, the whole business is undergoing the process of centralization that will perhaps assist in carrying the matter forward with more vigor and efficiency, if it be not attended by other consequences that will prove deleterious to local interests, and disastrous to the public welfare. This, however, remains to be seen.

Lands along the oil valleys were not originally held at very high prices. For merely agricultural purposes, many portions were perhaps as valuable as the land throughout the country; although, in many instances, they were utterly worthless, being broken or lying in very narrow valleys between precipitous hills. The lands along the Allegheny river, and along Oil and French creeks, that had been brought into a state of cultivation, were probably worth from ten to twelve dollars per acre previous to the development of the petroleum business. In other cases, where the land was uncultivated, and scarcely susceptible of cultivation, being of the character described by the pioneer, as "the more a person had the poorer he was supposed to be," the price could hardly be determined at all. In a few other cases still, the land had never been purchased from the State, being supposed utterly and incorrigibly worthless, and thus lying vacant. In this latter case the first individual who made the discovery of its orphan character located a warrant upon it, and for a trifle, embraced in charges, became the bona fide owner. These cases were rare, however, and small in extent, being usually but slender fringes bordering the streams, and bounded by steep and rugged hills. But in some cases, at least, they have proved valuable oil territory.

After the experiment of a general leasing of lands in the oil valleys had been in operation for some time, the original owners of the soil began to yield to what they considered tempting offers, and disposed of their lands. The prices paid, although considered large by the original proprietors, were very moderate when compared with the more modern rates. But as the business began to develop, prices increased, until the matter of speculating in oil lands has become a mania, and colossal fortunes have been made in this way.

The usual way was to obtain the price at which an individual landholder was willing to sell, and then to offer a certain fixed amount for the "refusal" of the property for twenty, thirty, or sixty days, as the parties might agree. The money was paid, and if "party of the second part" came back within the limited time, and paid the full price agreed upon, the deed was made out, and the property was his. If he did not come up to time, the sale was null and void, and the forfeit money belonged to "the party of the first part."

There purchasers were generally without capital; their whole stock in trade being their enterprise, and perseverance, and pursuit of the main chance. The confirmation of the purchase generally depended on the ability to make a further sale to a third party, of course, at figures greatly enlarged and multiplied. These lands during the past year have generally found their way into the hands of joint stock companies, that have of late become grand and ponderous institutions as connected with the oil business.

Many farmers, whose paternal acres were scattered along the oil valleys, have become suddenly wealthy. In former years they lived frugally and unostentatiously, and managed to make the product of each year supply its wants. They raised what grain was necessary for their own supply, and occasionally floated a raft down the Allegheny to Pittsburgh and the ports below, the product of which furnished them with other supplies. Thus the year rolled around, and there was no dream of future wealth. The riches of Venango county were supposed to be all developed. The last lingering castle had tumbled and vanished into thin air, as the iron business failed, and the last act of the drama was supposed to be on the stage. But many of the same farmers have become wealthy—some of them literally millionaires. One plain, worthy man, who had been born and brought up on the farm on which he was residing at the time of the new era, has recently sold his little possessions for one hundred thousand dollars. Another, on whose lands the business has been more fully developed, is about selling for half a million, and still another, with rather more productive lands, has the offer of two millions. And these are real bona fide sales, and the investment would be a good one for any capital that is seeking a profitable resting place.

But money, like everything else, seems to have a value that is merely relative. Many of these plain farmers do not appear to regard these high figures with any more favor than they did smaller ones years ago. One thousand dollars seemed to them a large sum years ago; one hundred thousand seems no larger now. It must be stated, for the credit of these men generally, that they have not been uplifted by the prosperity that has suddenly rained down upon them. Some have gone to other localities and purchased farms, and settled quietly down upon them; others are still in a transition state, and preparing for new changes.

## CHAPTER VIII.

### PRELIMINARY TO BORING.

In a popular cook-book there is this sage direction prefixed to a receipt for cooking a hare:—"First catch the hare." So in regard to the preliminaries to boring an oil well, the first thing is to obtain a site that will be at least promising. And in regard to this, taste and judgment are about the only guides. Doubtless there is something in the geological formation, but this, even to the scientific eye, is difficult to read. A broken and irregular formation of the rocks near the surface may be an indication of the condition of the rock far below; yet even this broken appearance near the surface may be owing to accidental causes that have not affected the underlying strata. On Pithole creek, where some new and almost startling developments have recently been made, the broken and shattered condition of the rocks would seem to be an index to the recent success. But this broken and fragmentary condition of the rock is not confined merely to the surface, as indicated by the pits and caverns that are found in its region, and from which the creek derives its name. It would appear, at least, to reach far down, and to be characteristic of that peculiar region.

At first "surface appearances," or the presence of oil in springs, or oozing through the surface of the ground, or manifesting its presence in the streams, was supposed to be an almost infallible indication of success. But the presence of this surface oil is not always a sure criterion in deciding upon a location for a well. Oftentimes very fine wells have been opened in localities where no oil has been found on the surface, and no appearance of oil having been obtained at any previous time in the neighborhood. There have been no more decided successes, from boring into the very ancient pits on Oil creek, already alluded to, than in many other places. In fact, wells have been bored in the bottom of these pits without the slightest success; whilst others, bored at a distance from them, where there was no appearance of ancient operations, have met with complete success. At a point on the Allegheny, about two miles above Franklin, there was a well-known oil spring forty years ago. It supplied the family that lived near it, as well as the surrounding neighborhood, with petroleum for medical and other purposes, to the extent of their wants. But for many years the supply has entirely failed. During a recent excavation, at the precise spot where it was known to exist at the time alluded to, for the purpose of laying the abutment of a bridge, no trace of oil was found—not even a discoloration of the soil. Yet this may perhaps be accounted for from natural causes. The little cavities, or perhaps mere seams in the rock, may have become closed, or the feeble supply diverted into other and more attractive channels. The very fact that it was found in minute quantities is evidence that it found its way through a very small orifice, or that it had come a great distance from the larger caverns from which it had been supplied.

In the earlier stages of operation, a ravine or sunken hollow, declining toward the river or creek, was generally selected as the best site, under the supposition, perhaps, that the character and appearance of the rock beneath was prefigured by the ground upon the surface. There is no doubt some geology in this, if not philosophy. But it is not a certain criterion. These ravines are generally not due to the force of circumstances beneath, but above. A stream of water may have originally worn the ravine for its bed, where now there is but the dry channel. Again, a stream of water or the efflux of a spring may have worn a furrow, or even a formidable ravine, as a channel for itself, where formerly there was a smooth, level surface.

Facts, too, are against this theory. In reality, facts have always proved themselves mortal enemies to any theories that have been set up in regard to the oil developments. Every operator could not obtain a ravine in which to plant his drill, because the applicants were numerous and the ravines limited in number. Many are obliged to content themselves with a smooth, level bit of ground as the theatre of the operation that was to render them, if not famous, at least wealthy. And in course of time it was found that the borer in the ravine, other things being equal, was no oftener successful than his neighbor. In fact, as theories have so frequently been shipwrecked and stranded in the progress of oil developments, it may even prove that, within the limits of the oil basin, or, perhaps, better, within the limits of the oil-bearing rock, that wells bored on the table land will be found equally successful with those on the river and creek bottoms. A fact or two has loomed up lately that points mysteriously in this direction.

In selecting a site, the proximity of neighboring wells has often been considered as of consequence. It is true one man's success is no great indication that his neighbor will succeed equally as well. Still it has its influence. Much more, the success of several in the same neighborhood will have a tendency to render a given piece of territory popular; it may, to a certain extent, demonstrate that in that particular region the strata beneath may be cavernous, and so the receptacle of the desired fluid to a greater extent than some other localites; but it is not infallible. A half-dozen wells may be yielding oil in a given locality, inducing the seventh comer to congratulate himself on the possession of a small piece of ground near by, or amongst the fortunate ones. But his drill may pass down, not into one of these unctuous caverns, but between them, and so go on down through the solid rock, until discouraged and disgusted he abandons the work.

On the other hand, a new adventurer may be successful, where others have failed, simply because his drill has penetrated a vein, or cavity of oil, whilst those who have preceded him have passed the vein, and found unbroken rock, until they considered the experiment fully tried. Besides new territory must be tested, and, in order to do this, there must be original navigators to open up the way.

As a general thing, wells in the same neighborhood do not interfere with each other. There are exceptions to this, however, but they are not at present sufficiently numerous to interfere with the rule. In rare instances the same crevice in the rock may be running on the same line with two wells; in such cases there may and will be an interference, unless the cavern or vein be sufficiently capacious to satisfy the demands of both. But in the great majority of instances the wells do not interfere, although not more than fifty feet asunder; so that fear need not enter into the calculation, when revolving the idea of locating a well in a given locality.

But not content with natural laws and natural manifestation, some seekers after the hidden wealth have resorted to supernatural means, and pretended supernatural agency. In their superstition and credulity they have invoked,

> "Black spirits and white,
> Blue spirits and gray,"

if not using the incantations and preparatory arrangements of the ancient sorceress, seeking, at least, to attain the same ends by more simple processes. Some have pretended to be in league with spirits who have pointed out to them the precise location where a successful well might be located, and even the depth to the luxuriant vein, that but awaited the tapping process, in order to a bountiful yield. But the spirits did not prove reliable. They were lying spirits, that led their deluded votaries astray, or rather the persons that professed to have "familiar spirits" were impostors, or deceived, and so their devices and practices came to nought.

There is a device that is sometimes resorted to, that may be mentioned for what it is worth, inasmuch as it has a small show of philosophy alleged in its favor. This is the use of the Hazel, or Peach tree rod, in order to point out the locality of the deposits of oil. It has been used by certain persons, "so long back that the memory of man runneth not to the contrary," in pursuit of water for the purpose of digging wells for domestic uses; and has had its believers and defenders.

The mode of procedure is to take a natural fork from the trees referred to, having its limbs of equal length and size, and to strip off the leaves, within an inch or two of the main stem. The extremities of these limbs are then grasped rather firmly in the closed hands with the backs downwards, being, at the same time, extended from the body. The supposition is that if there be oil underneath, the fork will revolve in the hands, having the two extremities for an axis; if not it will remain erect and immovable in the hands. The little bit of philosophy, and it is very small indeed, in the matter is, that there is a kind of magnetism by which the rod is disturbed and set in motion.

This is the theory. It has its advocates. It is no new thing under the sun. Works have been written on the subject, as applied to the discovery of minerals and mineral waters. It has had its advocates among scientific men in this country. "Silliman's Journal," certainly no mean authority in matters of science, published an article setting forth reasons for adopting the theory, and citing facts in proof of its correctness. In the Venango oil regions, as well as in that of West Virginia, it is asserted by candid men that very many instances are on record where rich wells have been indicated by this singular mode of manipulation. There may be philosophy in the theory, but it is difficult for the uninitiated to perceive it. Yet there are many things amid the hidden forces of nature that we cannot explain, and however strange they may seem to us, may not be safely discarded.

> "There are more things 'twixt heaven and earth, Horatio,
> Than are dreamed of in your philosophy."

Still if we had any advice to offer, it would be, not to trust too implicitly to the hazel tree theory, unless the site had other claims upon the attention.

In the earlier stages of oil operations the favorite locations, and, indeed, the only ones that were considered worthy of attention, were along the river and creek bottoms, and near the water's edge. Perhaps the idea was a complex one, having reference in part to the notion that the oil-courses would correspond with the water-courses, and in part to the matter of convenience in finding the rock near the surface of the earth, so requiring but little labor preliminary to boring. As to the first, it is an assumption altogether, that the rocky strata one hundred or a thousand feet beneath the surface is broken and cavernous, and adapted to the secretion of oil merely because an indentation has been worn in the soil to receive the current of a river or creek. It remains to be proved that the courses of streams are influenced in the slightest degree by the geological structure of the underlying rocky strata.

It has been asserted, however, that just at the foot of the steep bluff, along the course of Oil creek, there has been more uniform success than along the edge of the creek, and near the water.

It is very doubtful, too, about the correctness of the other branch of the idea, to wit: convenience in finding the rock more readily accessible on the beds of the streams than on higher ground. This would be assuming that the rock strata is to be found on a level over the face of the country, and that the eminences and hills were heaped upon it in different thicknesses according as they rise above this supposed level. There can be no doubt but that this supposed level is purely imaginary; nor can there be any doubt that many of the eminences of land are due to the upheaval of the rock beneath. Throughout the oil valleys, we find the rock cropping out from the sides of the hills, and rock too that possesses all the characteristics of that penetrated by the drill in the downward progress of boring. In the face of the steep, rocky bluff near Franklin, where an enterprising tradesman is excavating a cavern for the conservation and ripening of his "Lager Beer," the rock stratum appears mingled, and contorted, and interblended as though by some mighty convulsion of nature. The agency of heat, too, seems to have been superadded, as the material appears to have settled together from a plastic state. Still there is this slight advantage in selecting a site near the margin of the stream, that some of the superincumbent soil has generally been washed away by the action of the water.

Within the last year or two less attention has been paid to the low land, and many even prefer a high location. Quite as good success has attended operations back from the water courses, and wells are continually developing stores of wealth that have been bored some distance up the face of the hill. This is particularly the case along the valleys of Oil creek and Cherry run, where the strip of low land is narrow, and the territory is limited in extent. In some places the hill side is dotted with derricks and engines, and this not alone in the operation of boring, but in actual pumping.

The most startling development that has been made, in this phase of the question, has been on Pit Hole creek. A recent well has been opened that bids defiance to almost all the theories that have hitherto been promulgated. It is about six miles from the river, and on a piece of land that is at quite an elevation above the Allegheny. The well itself is of a less depth than the principal ones on Oil creek, and yet has pierced the fourth sand rock. Here is an argument in favor of high land as a site for boring; as well as proof that the surface of the country is no indication of the form of the underlying rock.

The question of sites for wells, then, is one not easily settled. For there is a vast difference between an oil well and a hole in the rock. In the latter there may be salt water, and other minerals, but petroleum is the great matter in question. New territory must be developed and new experiments tried, in order to enlarge the area of supply, and perhaps there are no general rules that can be laid down in the case. Proximity to other developed sites, is a matter that is not to be disregarded, and beyond this, strong faith, persistent effort, and a considerable amount of capital, are the best reasons on which to found hopes of success.

There is this, however, in the selection of a site for

boring a well: it would seem reasonable that near the base of the hills, rather than at a distance from them, a broken condition of rocks might be found beneath. If the hills be caused by upheaval of the strata underneath, then, in all probability, the underlying rocks will be displaced and shattered, and left in a disjointed cavernous state. And this is the condition most favorable to the secretion of oil. It is not at all likely that oil was formed near the surface of the earth, or where it is now found. It is formed elsewhere, and will generally be found in hollows, and cavities, and even large caverns. Something analogous to this condition of the rocks is found near the shores of our Northern lakes, and in portions of the rivers where the water runs rapidly. When the ice has obtained considerable thickness it is broken by the waves or current, and thrown up in great ridges, over which the water dashes and freezes in an apparently compact mass. But on examination it will be found to have an open structure with furrows, and crevices, and caverns.

Such is, no doubt, the condition of the rocks near the base of the hills, particularly if these hills have been formed by upheaval. Practice has often proved the correctness of this theory.

Even if there be a failure in a favorite locality, it is no very serious ground of discouragement. An inexperienced surgeon might thrust his lancet within a hair's breadth of the vein, and yet produce no blood; still it would be no evidence that there was no blood in the patient's arm. So a well might be sunk within a few inches of a fine vein of oil, and yet fail of making any discovery. An instance may here be cited to illustrate this idea. A well was in process of completion. The centre-bit had penetrated even below the region where oil was expected, with no tangible results. It was reamed out with the first reamer, with a similar result. But when the second reamer had passed down to the depth where oil had been expected, a copious vein of oil was opened, proving that the first reamer had passed along within an inch of the oil vein without disturbing it or giving any token of its presence.

As to the subterranean region which is to be explored, we can know but little, any further than the sand-pump of the borer brings the secrets of its prison-house to light. It is a long way down amid the rocky deep that the drill must force its way, before it reaches the region where the great Chemist has stored his treasure to await the necessities of his creatures. And nothing but the persistent genius of man, and this genius under the force of great pressure, could bring this treasure to light.

Usually there is, in this downward journey, a certain depth of earth to be passed through, varying from a few inches to fifty or sixty feet. After this a kind of gray or dark-colored shale or soap-stone, about two hundred and fifty feet in thickness, then a stratum of white sandstone, then shale again. These two species of rock alternate, until, as far as has yet been explored, there are four different strata of each. The third and fourth sand rocks, as far as we can judge now, appear to be the true strata containing unmixed petroleum.

This is the journey the drill must traverse through very hard rock, occasionally meeting on the way with a vein of fresh, but oftener salt, water; now and then a small vein of oil, called by the workmen "a show of oil," until the work is completed, and terminates in success or failure. The work is carried on in darkness, and far from the direct observation of the workmen, and yet with more precision and certainty than are the ordinary operations of the mechanic who has his work directly under his eye. When the well is completed the puncture may be a thousand feet deep, through nearly solid rock, and yet its calibre is doubtless as smooth and straight as that of a finished cannon throughout its entire length. If it were otherwise the work would be useless, and, in fact, could not be carried forward for any great length of time. If the well deflected from a straight line, the long augur-stem would soon become involved, and even if completed, the tubing could not be inserted.

The operator must prepare himself for difficulties in his downward progress. "It is not always May" in the process of sinking an oil well, any more than in many other operations in life. There are an hundred sources of accident that may interfere with even the practised borer, that no skill can at all times circumvent. Besides, from the great demand for operatives in this line, many who have no practical knowledge of the business are obliged to undertake it. The boring implements may break, the drill may become dĕtached from the augur-stem, and be left hundreds of feet in the dark abodes beneath, or the whole weight of the implements may stick fast, defying every attempt to get them loose. Some of these difficulties are almost constantly staring the bold miner in the face, testing his courage, putting his coolness to the trial, and calling forth all his resources of ingenuity, and judgment, and mechanical skill. All these contingencies should be looked at and anticipated, and even expected before the work is commenced. It is true, many a well is bored without any of these accidents happening. The entire work occasionally proceeds "merry as a marriage bell" to entire completion. So, sometimes, does a voyage across the ocean. But withal, the prudent sailor expects an occasional rough sea, and at times a fierce struggle with the elements, and is not disheartened when these things meet him in his rough experience.

There is another important consideration preparatory to boring—calculations must be made for total failure after all the toil and anxiety. There are more blossoms on the tree in spring than there will be fruit in autumn, and there are many more failures than successes even in the most promising portions of the oil region. An individual would hardly be justifiable in commencing with just sufficient capital to complete a single well, although it is not beyond the region of probability that the first well might be a perfect success. But, on the other hand, it might be a failure, and then all his hopes would be blighted. There would be more safety with capital sufficient to put down two wells; yet even in this case there would be considerable hazard. But if there should be capital sufficient to put down from three to half a dozen wells, then one success would remunerate for all subsequent failures, and leave a large margin of profit. The way to fortune through the oil regions, then, is not all plain sailing. There are difficulties and discouragements that must be met and endured. The two grand characteristics that warrant success are capital and perservering enterprise. With these success is certain, without them, failure is very probable.

# CHAPTER IX.

## MODE OF BORING.

We will suppose, then, that the site has been selected, either by the divining rod, by the geological features of the country, mere whim, or necessity, no other site offering; preparations are now made for boring the well. The little barque is to be launched that is to bear the operator to fortune or to shipwreck.

The boring of oil wells has become quite an institution in the oil valley. The first operatives were usually old salt-borers, who brought with them a wonderful amount of pretension, and were, of course, fully impressed with the dignity and professional importance of their occupation. The boring of a hole down through the earth, a distance of five hundred or a thousand feet, was a very important and responsible business. It required time and patience, as well as skill and judgment. Thus these old pioneers reasoned and acted. A few feet per day was all they expected, and were content with that. But soon a new set of men undertook the business, who run their drills with twice or thrice the rapidity of the pioneers, and made ten or fifteen feet daily in some varieties of rock. The old borers looked grave, shook their heads, and intimated that some dreadful calamity would ensue. The implements would break; the well would not be perpendicular nor cylindrical; everything, in fact, would go wrong. But the new spirit that had gotten under way triumphed. No greater liability to accidents was manifested than under the sober system of the old salt-miners. In this case, as in others, the spirit of innovation could not be stayed. To resist it was like "damming up the Nile with bullrushes." And now, after five years, the spirit of innovation, and the attempt at improvement in the processes of producing petroleum, is as strong as ever. Sometimes it is successful, at other times, it meets with failure.

The first thing now in order is the "derrick." This is a tall framework in the form of a truncated pyramid, about ten feet square at the bottom, and five at the top, having one of its four posts pierced with rounds at proper distances, to answer the purposes of a ladder, by means of which the workmen ascend and descend when necessary. The derrick is from thirty to forty feet in height, and has in the centre of its summit a pulley, by means of which the boring implements and chamber are drawn from the well. In fact, this pulley plays a most important part in all future operations, both in boring and pumping the well, should this latter process fortunately become necessary. In former years, the derrick was formed of four posts, sometimes rough as they came from the woods, sometimes hewn slightly, and connected together by as many cross-pieces as were necessary to give the structure strength and stability. Sometimes it was boarded up to furnish protection to the workmen, and at other times left open and uncomfortable. Latterly the derrick is made of plank, one and a half to two inches in thickness, spiked together at the corners, and braced transversely, to secure the proper amount of strength.

The derrick being finished, the next step is to commence the excavation, preparatory to boring. We will suppose that at the point selected, the underlying rock is from twenty to sixty feet beneath the surface of the ground. The drilling operation cannot commence until the rock is reached, and all danger of the caving in of the surrounding earth entirely removed. The usual practice is to dig a pit within the derrick, about six feet square, until the operation is interrupted by reaching the water. When the remaining distance is but slight, a wooden "conductor," as it is called, is driven down to the rock, having its upper end a few inches above the floor of the derrick. This conductor is sometimes made square, of strong oak plank, banded together with iron; sometimes it is a single log, bored after the manner of a pump-log. Through this conductor the boring is to proceed, as well as all future operations relating to pumping.

If the distance to the rock is great, instead of this wooden conductor, a strong cast-iron pipe is driven down through the earth, by means of a battering ram, attached to the derrick. This pipe has a calibre of about six inches, with walls of one inch in thickness. It is prepared in joints of about ten feet in length, which are connected together at the point of contact by wrought iron bands, about eight inches in width. When one joint has been driven down to the surface of the ground, this connecting band is heated to redness, a new joint is applied, and as the band cools a firm, close joint is formed. The battering ram is again put in motion, and the work proceeds until the rock is reached. As the blows are all in a perpendicular direction, the pipe usually goes down straight and in good condition. The derrick is now floored over, leaving the top of the pipe on a level with the floor, or an inch or two above it. The earth is then removed from the interior of the pipe, and the operation of boring is ready to be commenced.

Occasionally, however, this driving operation is interrupted by coming upon a huge boulder, or even a thin stratum of rock. This circumstance may sometimes deceive the workmen, particularly if they be new to the business. Yet this is not often the case; for, as a general rule, there is considerable regularity in the depth of the rock beneath the surface, and this depth is usually well understood in particular localities, and the careful workman need not be deceived. When one of these circumstances occurs, and the pipe, in its downward descent, gives notice of having met with an unwonted obstruction, the earth is removed from the inside of the pipe, and boring commences as though upon the real rock beneath. A hole is made through the boulder, or rock, nearly equal in size to the pipe itself, when the driving is resumed, and the pipe made to ream its way through the stone. Sometimes in these operations the pipe is broken, or driven out of its course. If it should descend upon the beveled side, or inclined face of the rock, it would most probably be forced out of a perpendicular direction, and the work would be effectually marred. When this is the case, the work must be abandoned, and a new location sought for, and the whole process commenced anew. Nor is it worth while to attempt saving the pipe that has been driven. It would resist successfully any attempt at drawing even by the force of the engine, whilst the presence of water would preclude the idea of digging it out.

If the workmen have been successful in driving the pipe safely to the rock, they are ready, after removing the earth and gravel from the pipe, to commence boring. The terms bore and boring are here used, because they are the popular ones. They are not strictly correct. Drilling would be a more correct word, as the process consists of a downward stroke from cutting and beating instruments, generally called bits. The operation, in obedience to popular use, will be here spoken of as boring.

The first thing to be described in connection with the boring apparatus is the cable. This is of sea-grass, usually from an inch and a quarter to an inch and a half in diameter, and in length as many feet as it is proposed to continue the well; this, however, is not material, as if too short it may be spliced, and if too long what is not needed may remain uncoiled upon the wheel, to be used in the boring of deeper wells.

One end of this cable is attached to the boring implements; it then passes over a pulley that is placed at the top of the derrick, exactly over the pipe leading to the well, after which it is coiled upon the shaft of the "bull wheel." This wheel plays an important part in the operation. It consists of a shaft about eight feet in length, around which the cable is to be coiled as it is drawn from the well. It also has a drum-head at one end, about four and a half feet in diameter, that serves for a band wheel, by which it is attached to the engine. The wheel is rigged in a horizontal position to the two corner posts of the derrick next to the engine. This portion of the machinery is used in removing the boring implements from, and lowering them into the well. It afterwards serves the same purposes with the tubing when the well is to be pumped. In withdrawing the implements, direct power is applied, either by the hand or by an engine; in returning them to the well a break is applied to prevent the descent from being too rapid.

In the operation of boring, the cable is the instrument most generally used as an attachment to the implements, as it is more convenient than any other device. Some parties, however, prefer rods, similar to those hereafter described as sucker rods in the process of pumping; these rods are about twenty feet in length, connected together by a screw and socket. The stroke of the drill may be a little more accurate and regular than with the cable; but the advantage is more than compensated in the use of the latter, by the ease and rapidity with which the implements are drawn out and returned to the well. In the use of rods, as each successive joint is drawn to the surface it must be unscrewed and set aside, until the whole are withdrawn. It is readily perceived that the disadvantage increases in proportion to the depth of the well.

We come now to speak of the boring implements generally used. They do not differ materially from those formerly used in sinking Artesian wells. As a general thing, bits of two sizes are used, the first and smallest of which only has a cutting edge. This is called the centre-bit. It is about three and half feet in length, with a shaft one and a half inches in diameter, terminating in a cutting edge from two and a half to three inches in width. This edge has an angle of about thirty degrees, and is, of course, of steel. Its upper end has a thread upon it by which it is screwed to the auger-stem when ready for use.

The other bit is called a "reamer," and differs from the former in having a blunt instead of cutting edge. The reamer is about two and a half feet in length,

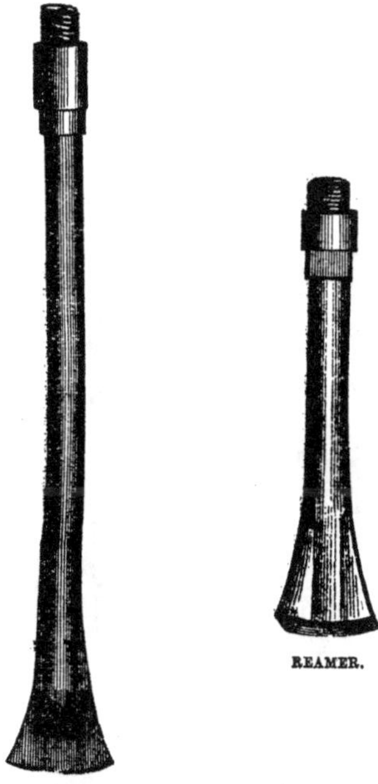

CENTRE-BIT.

REAMER.

with a shank two and a half inches in diameter, terminating in a blunt extremity from three and a half to four and a half inches wide by two inches in thickness. It is connected to the auger-stem in the same manner as the centre-bit. A large reamer weighs about fifty pounds.

The auger-stem is a simple shaft generally about twenty or twenty-two feet in length, and about two and three-quarter inches in diameter. A section of it is seen in the engraving. It is designed by its weight to give power and force to the bits attached to it.

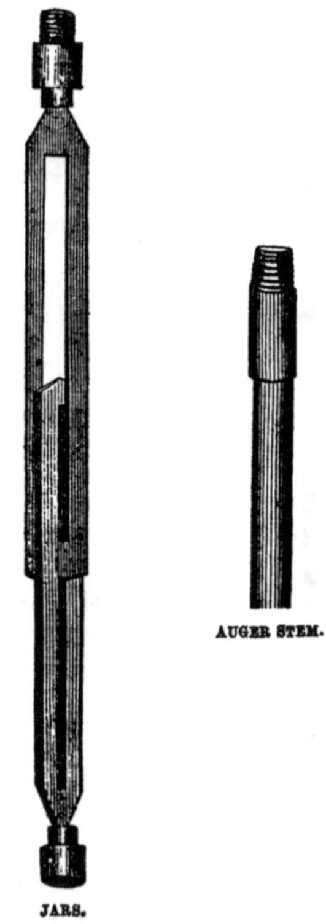

AUGER STEM.

JARS.

Connected with this auger-stem above is an important arrangement called, technically, "jars," made of two elongated links, or loops of iron, hooking in each other, one connected to the auger-stem, the other to the cable above. This arrangement serves to jar the bit loose when it has a disposition to stick fast in the well. The operation is readily seen, as it is noticed that at each stroke of the bit in the well, the upper links in the jars may sink down a foot or more, and as the power is applied it rises against the upper end of the lower jar with a stroke like that of a sledge hammer, and thus will overbalance any ordinary disposition on the part of the bit to adhere to the walls of the well. To the upper end of the jars another bar of iron, about six or eight feet in length, is attached called a "sinker." To the top of this the cable is attached by an arrangement called a "rope-socket,"as shown in the engraving.

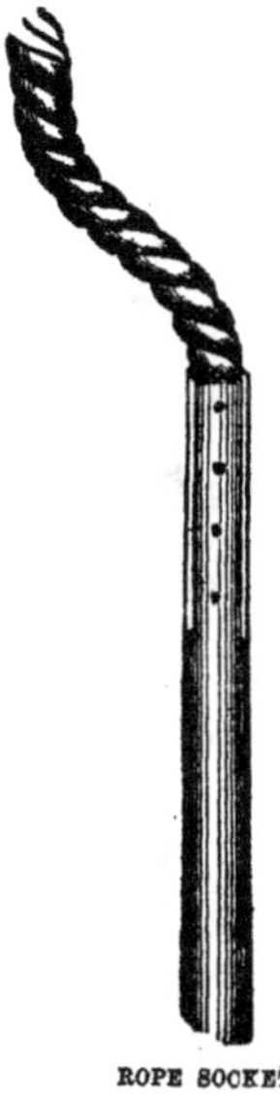

ROPE SOCKET.

Thus far as to the implements at the extremity of the cable, and that work in the well. Of course, as the work proceeds, and the bits descend, the cable must be continually lengthened, and this in proportion to the rapidity of the work as modified by the nature of the rock. This operation is provided for by an arrangement called a "temper-screw." It is a simple screw with rather coarse threads with head downwards, and working in a narrow iron frame that is attached above to the "walking-beam" of the engine. To the head of this screw is attached a transverse bar of iron some four inches in length, that serves as a handle by which the cable is lengthened at the pleasure of the workman. When the same is lowered to its entire length, it is detached from the cable, and screwed through the frame to its entire length, and is ready for operation again. To the lower end of this screw is attached a peculiar kind of iron clamp that embraces the rope, and, by means of a side screw, holds it firmly and securely. When the temper-screw is worked upwards through its frame, this clamp takes a new position on the rope, lengthening it by the length of the screw.

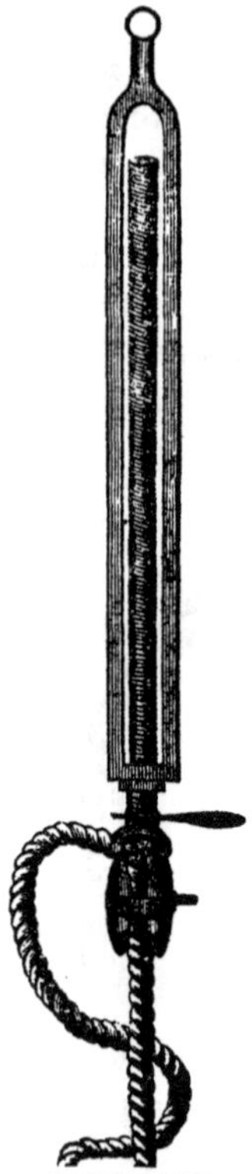

TEMPER SCREW.

The entire weight of the implements attached to the cable is from six to twelve hundred pounds. The motion is simply a perpendicular one, although the bit is constantly turning. The power applied, whether by hand or engine, is applied to raising the bit, the cutting force is all imparted by the weight of the implements, falling, when relieved of the power that raises them, either in the downward spring of the pole, or the downward motion of the walking-beam of the engine.

The power applied is now to be considered. In the earlier stages of the business, it was very often that of two or three men working at a spring-pole; at present it is universally a steam-engine. The spring-pole operation was a severe one on the workmen, and rather tedious; but for the first hundred feet operated very well. The spring-pole consisted of a green sapling about forty feet in length and ten inches in diameter, with the butt end made fast in the ground, or attached to an upright pole; a second post, ten or twelve feet from the butt, acted as a fulcrum, while the pole passed over the well, and about ten feet above it. The implements of boring were attached to this pole, and the power adjusted near its smaller extremity. This was applied by the strength of two men bearing down upon the pole. Sometimes a small stage four feet square was hinged by one of its sides to the side of the derrick, and the other side suspended to the pole. In this case the two men stood upon the stage, and brought down the pole by throwing their weight upon the side attached to it, and permitted it to rise by throwing their weight on the side next the derrick. In either case the spring of the pole brought up the implements, whilst the motion of the pole downwards permitted the stroke. In boring by hand, the implements are not so heavy as those designed to be operated by an engine.

Another mode of operation was sometimes used in the earlier stages of the business, in which a chain was used, causing the most unpleasant and outrageous clangor that can possibly be conceived. From its unpleasant and horrible noise and associations, it was called by the urchins the "chain-gang system." It was a laborious process, and a terror to the whole neighborhood where it was employed, and soon fell into disuse, much to the satisfaction of persons residing within half a mile of the scene of operations.

There was still another system by which human muscle was brought into requisition. This was named by the same astute authority quoted above, the "kicking pony system." A short, elastic ash pole, of ten feet in length, was arranged over the well, working over a fulcrum, to which was attached stirrups, in which two or three men placed each a foot, and by a kind of kicking process brought down the pole, and produced the motion necessary to work the bit. By this process the strokes were very rapid, generally making two in a second. It was adapted, however, only to shallow wells, and for a distance of an hundred feet or less answered a very good purpose. It, too, has been laid to rest, having performed its work.

In many instances horse-power has been used to good purpose. It had this advantage certainly, that it relieved human muscle of much drudgery. There were different patterns of horse-powers. Sometimes one horse did the work, and sometimes two or three. In their general features they resembled the horse-powers of threshing machines, the horses walking around the centre, and over a kind of tumbling-shaft that afforded the perpendicular motion necessary to raise and depress the bit.

Occasionally water-power has been used to advantage, both in boring and pumping. This is particularly the case in the neighborhood of water-power already established. A very simple machinery will enable the borer to carry on his operations in connection with those already going forward. In the vicinity of small streams, too, a dam can be thrown across, and the water, that had else gone idly by, be pressed into the service, and made instrumental in relieving brawny arms of the severest drudgery.

Still, all these processes, however useful in their time and way, have given place to the steam-engine. They have done their work, and often did it well, and were adapted to the exigencies of the times. Men of small capital, and companies of small resources, were not able to purchase engines at the first; and, indeed, many of the first operations in boring were mere experiments, intended to develop the resources of the country, and it was thought hardly worth while to incur the expense of an engine on such slender expectations. In addition to this, no one, unless the veriest dreamer, supposed it would be necessary to go beyond one or two hundred feet.

But all these arrangements have now been discarded, and the engine has taken their place. These engines vary in pattern, but the general principles are, of course, the same. The only material difference in form even, is, that some are stationary, being set in masonry, while others, and the largest portion, are portable. The first engine that was brought to Franklin for oil purposes was a portable one from the establishment of A. N. Wood & Co., of Eaton, Madison County, N. Y. This establishment was the first in the field, and amidst all the rivalry that has sprung up it has steadily held its own, and maintains its position. Other establishments are doing nobly, turning out the finest work, and manifesting a spirit of accommodation and interest in all that pertains to the prosperity of the oil region.

The pioneer engine of Mr. Wood was a four horse portable one. It was then considered sufficiently powerful for the purpose, but the experience of five years has shown that engines of from eight to ten horse-power are best adapted to the purpose; particularly in pumping deep wells, where a vast amount of power is necessary to overcome all the obstacles in the way.

In these portable concerns, the engine is built upon

the side or top of the boiler, somewhat on the plan of a locomotive. Underneath the boiler are feet, that are screwed to a light wooden frame, and the whole apparatus can be readily transported from place to place, and set in operation as soon as it can be set level upon the ground. They occupy a very small space, too, as they are built on the tubular principle, and in a small space present a large amount of water surface to the action of the fire. Sometimes these engines are built

WOOD'S PORTABLE STEAM-ENGINE.

upon iron wheels, so that they can be transferred from one point to another without the trouble of putting them upon a wagon. This arrangement, however, must be rather suggestive of failures in boring to the workmen, as the sanguine oil man would prefer not moving his apparatus until actually worn out in the work of pumping oil.

In adjusting the engine to the work assigned it, but little machinery is necessary. Sometimes a band-wheel, *c*, with a crank, is geared to the engine, to which a pitman is attached, the upper end of which is attached to one extremity of the "walking-beam, *a*, while the boring implements are attached to the other. A rude house is erected for the engine, into which one end of the walking-beam extends to receive its attachment to

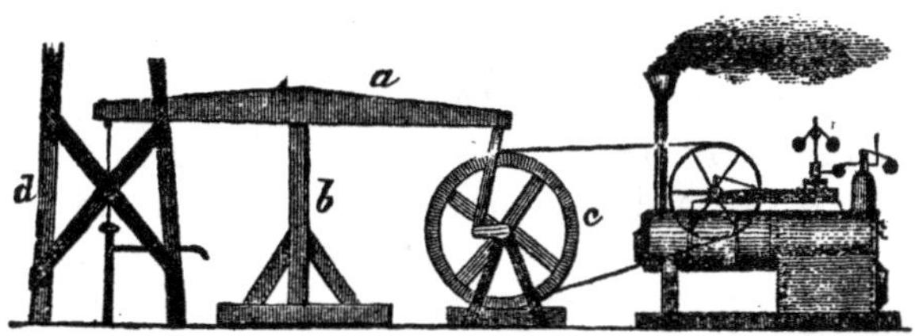

SECTION OF A PUMPING WELL EXHIBITING THE APPARATUS.

the engine. The other end of this walking-beam reaches into the derrick, *d*, a portion of which is shown, and terminates just over the well, raising and depressing the implements used in boring and pumping.

The walking-beam itself is made of wood, generally sixteen or twenty feet in length, about one foot through in the middle, and diminishing gradually to the extremities. It is balanced upon a post eight or ten feet in height, called a "Samson post" (*b*), and moves freely upon a pivot. The Samson post is set firmly in the ground, or framed into a horizontal beam that lies in the ground.

We are now ready to commence the actual work of boring, having the iron pipe driven to the rock, the bit arranged and engine attached. The workman seats himself just over the well, on a slight stool about three feet in height. The bit is let down through the pipe or conductor until it rests upon the rock. The rope attached to it passes over the pulley at the top of the derrick, immediately overhead, and descends and is coiled around the shaft of the bull-wheel. The end of the walking-beam in the derrick is now loosened, and the clamp of the temper-screw made fast to the rope, and the operation is ready to commence. The throttle-valve of the engine is thrown open, the walking-beam begins its vibrations, the bit rises and falls, and there is that sharp, clinking sound so familiar to the workman in the first stages of the operation. The workman's hand rests upon the transverse iron handle attached to the temper-screw, just above the rope, and as it descends, turns the rope, and with it the bit or drill, partially around, so that each stroke of the bit on the rock beneath is slightly across the cut that preceded it. And so the work goes on in darkness and under water, and still with great precision and certainty.

After the operation has proceeded about two feet, the work begins to flag. The bit becomes dull, and the well begins to clog with sand; and in addition, the temper-screw has nearly run out, in lowering the rope for the advance of the drill. The clamp is now loosed from the

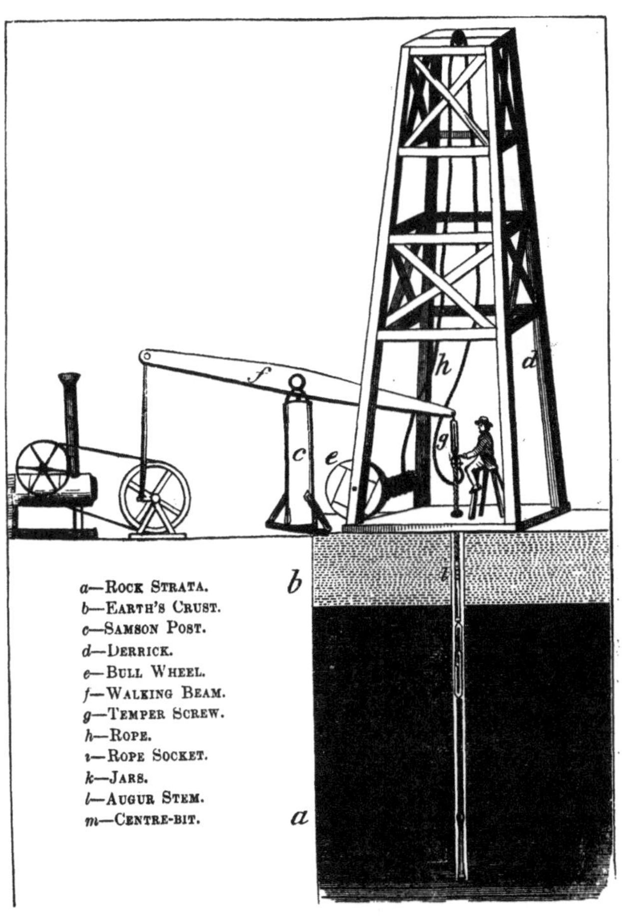


Section of a Well in process of boring.—*Page* 112.

the rope, and by means of the bull-wheel, either by the instrumentality of human strength or attachment to the band-wheel of the engine, the implements are drawn from the well. The centre-bit is detached from the auger-stem, and one of the reamers attached in its stead. This reamer has a blunt face, and is perhaps one inch wider than the bit that preceded it. Its office is to enlarge the hole, by the sheer force of blows. After examining carefully, to see that all the arrangements are in good condition, the reamer is lowered into the well, sinking by its own weight, and eased down by a break applied to the bull-wheel. The work is resumed. The fragments of rock that are cut and broken away descend to the bottom of the well in the form of sand. When the reamer has done its work it, too, is withdrawn, and the sand and water mingled into a batter is drawn out by means of the "sand-pump." This is a simple copper tube, about six feet in length, with a diameter a little less than that of the well, and furnished at the lower end with a simple valve opening upwards in the interior of the pump. At the top it is surmounted by an iron handle, to which a small rope is attached. At the bottom the sand-pump is

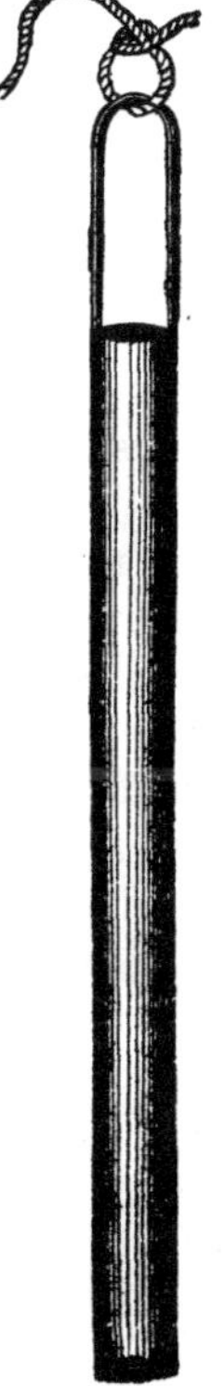

SAND PUMP

weighted with lead, to assist in sinking, as well as in performing its work when below. The pump is let down by this rope, and when at the bottom is agitated by the hand of the workman for a few moments, when the sand is forced up through the valve, and thus drawn from the well. This operation is usually repeated once or twice, until nearly all the detritus is removed from the well. The operation is then continued by putting down the other reamer, and completing the work as far as it goes. Many workmen, however, use but one reamer, completing the work in this way that has been commenced with the centre-bit.

The frequency with which the drill is drawn from the well depends on the nature of the rock. This operation is more frequent in hard sand-rock than in shale or soap-stone;—in the former about every foot or two; in the latter it can sometimes be run two and a half or three feet.

The bit must frequently be sharpened or dressed. This is done nearly every time they are drawn from the well. Duplicates are usually provided, so that no detention may be necessary in the operation. This dressing is performed by the borer himself; and as very few implements are necessary, a simple miniature smith-shop is connected with the well, containing a forge and bellows, with anvil and hammers. The centre-bit is sharpened without any very great care being taken as to preserving its exact size. But with the reamers the greatest care must be taken to preserve the exact width, else they would be liable to get fast in the well. To provide for this a gauge is used, by which they are brought to the exact size, pains being taken to apply the gauge, at the same temperature on each occasion.

In the matter of tempering great care is to be taken. If too soft, the bit will not be sufficient for the work; the cutting edge will be turned, and the blunt surface battered and fail of accomplishing the object. On the other hand, if they are too hard the edge will be broken, and the reamer probably break and leave a portion of its steel in the well, to be a source of trouble and annoyance in carrying on the work. The practice of some, and their experience has been favorable to the plan, is to bring the bit, when properly dressed, to a fine cherry-red color, and then to plunge it in water until it arrives at the proper tinge of blue. The reamer having been brought to the same color in the fire, is suspended by a rope over a bucket of water, permitting the face of the reamer to extend into the water about one-fourth of an inch, and suffering it to cool gradually. In this way the face or working portion of the bit is brought to the proper degree of hardness for the work assigned it, whilst the shank and part immediately adjoining the face are so annealed as to guard against fracture, in the repeated blows that accompany the operation of drilling.

As a general rule, the well is full of water almost to the surface of the ground. This, at first, results from the surface water flowing in from above, and afterwards from small veins of water that are pierced on the way downwards. This water answers an admirable purpose in carrying on the work. Were it not for its presence, the sand and debris of the rock set free by the drill would clog at the bottom of the well, and become as hard almost as when in its original position; but the water reduces it to a fluid, muddy mixture, that permits the drill to reach the rock, and at the same time presents it in good condition to be taken up and withdrawn in the sand pump.

In the journey through the rocks, there is not the monotony that might be supposed. Although it is a pathway through a sandy Sahara, yet, as in Africa, there is an occasional stream of water passed on the way, that adds variety to the work, and interest in its progress. These water-veins are found at all distances from the surface, and in the sand rock as well as in the dark shale. Nor is this mere imagination. The workmen has the practical evidence. Sometimes, after having worked the usual length of time, and bored the usual distance, he withdraws his drill from the well, and, to his surprise, finds all the implements clean and bright as though carefully washed in clean water. On applying the sand pump, he finds neither sand nor muddy water as usual; indicating that a vein of water has been tapped that has carried away all the sand and mud that had accumulated in the two or three hours' work. As the work proceeds, veins of salt water are always met with. Not a well has been bored to the depth of two hundred feet without meeting with this saline rock, or, at least, veins of very strong salt water. What weight this fact may eventually have upon the theories as to the origin of petroleum cannot now be determined. The fact is plain as to its invariable presence in all boring operations. In some wells bored in Franklin, old salt miners assert that the salt water thrown out from a two and a half inch tube would condense into thirty barrels of salt per day.

During the progress of the work there is more or less carburetted hydrogen gas set free. This is first noticed in the sand pump, as it is brought to the surface with its deposit of sand. Minute globules rise to the surface, lively and sparkling, like a glass of soda water. Sometimes this supply is so abundant as to cause an ebullition on the water of the well at its surface, resembling the furious boiling of a pot. This exhibition of gas was formerly considered a very favorable indication of oil, yet is by no means infallible. Still the eager workman is always on the alert for accompanying symptoms, and a flow of gas, or better still a vein of oil is hailed with as much satisfaction as were the joints of cane and banana leaves, floating on the water, to Columbus and his crew, as they approached the shores of the new world. Yet many another navigator situated as Columbus was, would have given over the voyage when almost in sight of land, with all these indications of success; or would have changed the course of his bark so as to have missed the land after all his anxiety and toil. So now many an oil seeker, with both gas and small veins of oil, wearies and becomes discouraged almost in reach of a fine vein of oil, as proved by the success of succeeding workmen in the same well; or perhaps more frequently he bores past the vein, leaving it but a few feet or inches to the right or left of his drill.

In passing through these oil veins their presence is indicated by the oil rising in the sand-pump, and floating upon the surface of the well. Sometimes they continue to manifest their presence while the work progresses; sometimes they disappear altogether, as in cases where a large vein of water is passed that carries the oil with it from the well. In the earlier stages of the business, this "show of oil," as it is termed, was considered most favorable to ultimate success; but latterly it is not regarded as essential, as many first class wells have been discovered without the intermediate show; and, on the other hand, there has been many a brilliant show that has resulted in failure and disappointment.

A strange feature has been discovered in the process of deep boring, particularly in the Oil creek valley. This is what is termed a mud vein. It appears to be a thin stratum of mud or clay of a most tenacious character, very annoying to the workmen, and sometimes most disastrous in its influences upon his work. This stratum is found at a depth of about five hundred feet from the surface, and is usually about five inches in thickness.

As the work advances, a register is kept by the judicious borer of the different strata passed through, and also of the veins of water and soil met with, in order to the formation of an intelligent judgment in the matter of tubing the well. This, it will be at once perceived, is of the utmost importance. If the well is a success, it will be necessary so to arrange the tubing that all the surface water, and that which proceeds from internal veins, shall be entirely excluded. It is likewise necessary that the point selected for shutting off the water from above be in the smooth solid rock, else the tubing will be imperfect, and pumping a failure. This point can be ascertained from the register.

In connection with this register, samples of the rock at different depths are often preserved. This is done by putting a small part of the contents of the sand pump upon a board, and noting the depth from which it was obtained. In this way a tolerably clear opinion can be formed as to the nature of the different strata passed through. If it were possible to pass down by removing the rock in solid core, as has been proposed, such an opinion could be much more correctly and intelligently formed.

But, as might be readily supposed, this operation of descending a thousand feet amid the rock is not without its troubles and discouragements. There is many a "hill difficulty" on the journey that must be ascended, albeit the way is downward. Sometimes these difficulties are sufficient to appal the strongest heart, and wear out the most persistent energy. Sometimes they arise from want of caution in the workman at the rope, sometimes from an imperfection in dressing and tempering the drill, and sometimes, perhaps oftenest, from circumstances that can neither be foreseen nor avoided.

Sometimes a bit breaks, leaving a piece of hardened steel deep in the recesses of the rock. When the fragment is small it is often beaten into the sides of the well, and disappears forever, without much annoyance. When it is larger the difficulty is great, for it may not be possible either to beat it to pieces or extract it from its bed. Sometimes the bit becomes detached from the auger-stem by reason of the loosening of the screw from its socket. This difficulty may be aggravated from the fact that the workmen may not be aware of the displacement, and for hours be beating upon it with the top of the auger-stem. When this happens various plans are adopted to extract the truant drill. Sometimes in a neighborhood as many tools accumulate, designed for such accidents, as there are instruments in a surgeon's office. There are persons, too, who get a reputation for such skill in extracting implements from wells, that they are sent for from considerable distances, and demand extravagant prices for their services. These instruments are of various forms, at the lower extremity; but are attached to the auger-stem or sinker in the same manner as the drill itself. If the bit is standing perpendicularly at the bottom of the well, the case is soon disposed of. An instrument with a spring socket is let down over it, which lays hold upon the protuberance, just below the thread, and brings it to the surface. If the drill has been beaten so that its upper end has been driven into the rock at the side of the well, the case becomes complicated; and some other remedy must be applied, and a different instrument adopted. Oftentimes weeks and months are consumed in trying different plans, and adopting different remedies, until the patience and resources of the proprietor are exhausted, and the well is abandoned in despair.

Again, the drill will sometimes get fast in the well without any displacement; ordinarily this difficulty can be overcome by repeated blows through the medium of the jars. Lately a difficulty has arisen from the mud vein, already alluded to, that is in the highest degree annoying and perplexing to the borer. The mud will suddenly flow into the well, and settle around the drill almost as firmly as the rock itself. When boring on it or below it, the experienced workman, when about to withdraw his drill will have a hand at the bull wheel, and the instant the walking-beam ceases its motion a turn or two will be taken on the wheel so as to raise the bit above the mud, as it sets almost as quickly as plaster of Paris. Sometimes, however, this mud will flow in, and fill up the well for the depth of twenty feet, burying up the implements even above the jars. This renders the jars useless as a propulsive power. The workman will now resort to a desperate experiment. Attaching a knife-blade to a series of poles, he will manage to cut off the cable just above its attachment to the sinker, and, then withdrawing it, attach to the pole a spear-pointed instrument with which the clog around the implements is effectually probed, when an extra pair of jars are attached, and an effort made to jar the unfortunate tools from their place of burial. It is marvellous what success often attends the effort to extract tools that are often four or six hundred feet below the surface. But this success does not always crown the labors of the indefatigable borer. Occasionally, after spending almost as much time and labor as would suffice to bore a new well, disappointment attends every effort, and it must be abandoned at last. All over the oil region wells may be found that have been abandoned, leaving from fifty to nine hundred pounds of iron and steel deep in their rocky cells.

As the workmen proceeds, usually the volume of gas increases as he approaches, what some denominate the "oil-bearing rock," but which, perhaps, might be more appropriately named the oil-surrounding rock. In the case of flowing wells, when the vein of petroleum is reached, the gas rushes forth with such violence, and the upward pressure is so furious as to force the implements from the well, driving them through the derrick, in their resistless fury.

The depth to which a well is sunk depends on many circumstances. Sometimes it is regulated by the depth of wells in the neighborhood that have proved successful, and sometimes by the "show of oil" in the well. Even when a good vein of oil has been pierced, and the desirable fluid has been found in large quantities, the well is still sunk many feet deeper, in order to provide a receptacle for sand and particles of earthy matter that fall from the sides of the well, and thus prevent them from clogging up the vein.

But the presence of oil in a well is no infallible evidence that the work is a success. It may all be from a few minute veins, that will soon exhaust themselves without the aid of the pump. Neither is the absence of oil, even its total lack, any infallible indication that the well is a failure. The volume of water from above and around may be keeping it back, until removed by the pump.

In the beginning of oil operations, the curiosity of the public was so great as to be a source of serious annoyance to the workmen. Crowds thronged around the derricks, all taking a great interest in the work, eager to become acquainted with its progress, and to learn of its ultimate success. Several times in the course of an hour the man seated upon the boring-stool, would be interrogated as to the depth he had reached, whether hard or soft rock, how he got clear of the sand set free in boring, whether he had found much oil, or how deep he designed to go before giving up. To all these queries and an hundred others, reasonable and unreasonable, the good-natured borer was obliged to make some reply. Sometimes the answers were correctly made and sometimes not, just as the mood might be upon him.

Occasionally a new device was resorted to. A flaming poster was prepared and put up in a conspicuous place outside the derrick, on which the whole state of the case was set forth, embodying alike facts, results, hopes, and assurances. Perhaps it would set forth the following, or like scheme:—"We are three hundred feet deep in the sand rock. We have a good show of oil. We make ten feet per day. We expect to have a good well. We have no fear of failure. We are employed and paid for boring this well, and not for keeping a general intelligence office!"

Perhaps one of the most suggestive intimations was found painted on a board, and affixed to the outside of the derrick, where patience certainly had room to have her perfect work. The men had worked long and persistently without "a smell of oil," as the workmen term it, and were still nearing the centre of the earth at the rate of ten feet per day. Their inscription was "OIL OR CHINA," in blazing capitals. Here was the persistent determination of the people where every man is a sovereign; the purpose must be accomplished even if pursued to the other side of the globe.

In this labor the progress depends on the nature of the rock; in the shale it is much more rapid than in the sand rock. Perhaps, on an average, about ten feet per day is fair progress. At many establishments the labor is continued night and day, employing two sets of hands —one set commencing at noon and continuing until midnight, the other from midnight until noon. A well six hundred feet deep will require from five to ten weeks to accomplish the work, according as the work is carried on twelve or twenty-four hours per day.

The cost of sinking an oil well to the depth of

six hundred feet may be learned from the following estimate:—

| | |
|---|---|
| Forty feet of metal pipe at $6 per foot | $240 00 |
| One engine, ten horse-power | 1,600 00 |
| Band-wheel and belting | 125 00 |
| One set of boring implements | 325 00 |
| Derrick complete, with bull-wheel, walking-beam, and samson-post | 100 00 |
| Six hundred feet of cable and sand-pump rope | 100 00 |
| Drilling six hundred feet, at $2 50 per foot | 1,500 00 |
| Six hundred bushels of coal, at 50 cents | 300 00 |
| Total | $4,290 00 |

To all these a small amount may be added for contingencies. With this view a very near approximation may be made to the expense of boring an oil well.

In the matter of boring, as in other branches of business, new plans have been suggested, and new machines brought forward, with more or less success. Besides the implements described at length in this chapter, there has been brought to notice what is called the Z bit, from its resemblance on its cutting face to two letters of this variety crossing each other at right angles at the middle of the stem. In order to obtain a proper idea, however, the transverse portion of the letter must be imagined to be quite long, and the parallel portions quite short. The whole face of the bit corresponding with the different portions of the letter is furnished with a cutting edge. The body of the bit resembles somewhat a blacksmith's mandrel, and is quite heavy. It is attached to a cable, and operated as the utensils already described. The idea of this bit was brought from California, but it has not yet had sufficient trial to warrant an opinion being formed as to its merits or demerits.

Another instrument has lately been introduced, more novel in its principle than any that has preceded it. It is called the "diamond drill," and is said to have been patented originally in France, by Rudolf Leschand, and introduced to the United States through the "Scientific American Patent Agency." The drill consists of a long thin tube or cylinder of steel, in the lower end of which fifteen diamonds are set around the circumference. These diamonds are really the cutting instruments. The drill is operated by machinery from above, that gives it a rotary motion of very great velocity. The rock is cut out in a solid core, and is removed in pieces by clamps let down from above. By this process the patentee expects to put down a well of four hundred feet in two or three weeks. If this plan of boring is successful, it will be a valuable improvement in the matter of boring, both as regards expense and time, as well as assist in elucidating many interesting questions that arise in regard to the rock, the formation of strata, the frequency of seams and small apertures, as well as other matters both curious and valuable. This drill, like the one previously mentioned, has not been in use a sufficient length of time to test its value and importance.

## CHAPTER X.

### TUBING AND PUMPING.

After the boring has been successfully accomplished, the next operation is testing the well. In other words, it is to decide whether the result of two months' labor, and the expenditure of several thousand dollars, is simply a deep, worthless hole in the rock, filled with salt water, or a bona fide oil well, from which is to flow wealth, and comfort, and ease. The operation of testing, it may be supposed, is like that of the merchant as he balances his books at the close of the year, to ascertain whether he is wealthy or simply a bankrupt. No wonder the operator holds his breath as the engine is put in operation, and the first strokes of the pump begin to exhaust the chamber. But this work of testing is not the labor of a day simply. It may be prolonged over many weeks and months even. No doubt many a vein of oil is pierced that is lost to the owner of the well, from the simple fact that the tubing has been defective, and the means used for exhausting the water inadequate. In the process of boring, too, a valuable vein of oil may be closed by the beating and pounding process of reaming, effectually closing the aperture, as though by the use of oakum and caulking iron, with a design of excluding the oil.

The introduction of a tube and pump is the only test

that can be depended on in deciding as to the success or failure of a well. There may be a rich stratum of oil on the top of the well, and on the surface of the sand-pump, and yet the work prove a failure; for all this oil may be from minute veins, that will yield but little in the aggregate. Again, there may be no surface show whatever, and yet the result be a complete success. A vein of water may be pierced that leads to an oil cavity, it may be, far away, and it requires but to have this water vein exhausted in order to set free the oil that is imprisoned in the rocky bed. We have found the well full to the surface during all the stages of boring; this must be overcome, and a vacuum formed low down, and as near the cavity containing the oil as possible.

Tubing consists in putting a continuous pipe down to the oil vein, and then excluding the water from above. This tube is denominated the "chamber" by the oil men, and varies from two to three inches in diameter, consisting of sections about twelve to fourteen feet in length, connected together by screw and socket joints. The sections are put together as the chamber descends into the well, and detached when it is taken out, as circumstances render this operation necessary.

At the commencement of oil operations this chamber was almost universally made of copper. It was supposed that no other material would successfully resist the action of the salt water that is found in every oil well. It had, moreover, the advantage of being light and easily handled. But it was very expensive, and as prices advanced, required a small fortune to purchase a pump. It lacked also in strength, even in wells that were not very deep, frequently bursting under pressure, and exposing the workmen to tedious and perplexing delays.

But when deep boring was resorted to, and wells of from five to eight hundred feet were to be tubed and pumped, the use of a copper chamber became impracticable.

The use of iron as a material for the chamber was then resorted to with almost complete success. It has not suffered from its saline bath as much as was expected. It is, as a general thing, shut off from contact with the atmosphere, and being always coated with oil, the salt water makes but little impression upon it. This tubing is of wrought iron, of sufficient thickness to insure strength, and connected together as the copper chamber was, with screw and socket joints. The size of the chamber depends on the supposed capacity of the well, being generally, for pumping wells, from two to three inches in diameter.

GAS TONGS.

This chamber is put down into the well, a new joint being added as each preceding one sinks down to the level of the derrick floor, until the lower end reaches the neighborhood of the oil vein. The joints are screwed together and detached by the use of a peculiar kind of tongs, that clasp the chamber firmly, without at the same time injuring it. The register kept during the process of boring enables the workman, as a general thing, to determine where the vein of oil is to be found, and so to arrange the proper length of the chamber.

But there are usually many veins of water passed through in the process of boring, as well as a heavy body of surface water pressing into the well, and some device must be resorted to in order to shut off this water from the oil vein, and thus produce a vacuum. This is accomplished by applying what is called a "seed-bag" to the tube at the point where this shutting off would be desirable. This point, too, is regulated by the register, as it is important that it be in smooth, solid rock.

The seed-bag is a tube of strong leather, about eighteen inches in length, and of a diameter somewhat larger than that of the well. This leathern cylinder, or pipe, is put around the metallic tube at the proper point, and firmly tied at the lower end. From a pint to a quart of flaxseed is then poured in, and the upper end tied rather more slightly than the lower. When the chamber is sunk to its place in the well, the seed in this leathern receptacle swells so that in a few hours the bag distends and effectually shuts off all water from above. When it is necessary to withdraw the tubing from the well, the effort of raising it will break the slight fastening at the upper end of the leathern sack, permitting the seed to escape, turn the bag wrong side outwards, and leave it in proper condition for filling, and readjusting when the tube is returned to the well.

Other devices have been proposed to accomplish the same purpose as the seed-bag, but as yet nothing has succeeded so well as the old-fashioned arrangement of the leathern sack filled with flaxseed. It is simple, it is practical, it accomplishes the purpose. Other methods that have been proposed are cumbrous, difficult in their adjustment, and so far have not succeeded in fulfilling the expectations that were entertained in regard to them.

The well being successfully tubed, if it be a flowing

well, the gas and oil will flow forth without any farther trouble from the engine, or any other power. If there is not sufficient gas to force the oil up, preparations must be made for pumping.

For wells that are to be pumped, a pump-barrel is placed at the lower end of the tube, to be worked by a piston. The valves at the lower end of this pump-barrel are of various kinds, sometimes working with a hinge, and sometimes as a detached ball. To the piston in the pump are attached "sucker-rods," reaching to the top, and attached to the walking-beam used in boring the well. These sucker-rods are of wood, in joints about twenty feet long, and connected together by iron sockets and screws, and are connected together and detached as they are put into the well or taken out. When taken out, both the chamber and sucker-rods are set on end in one corner of the derrick; and in their operation the bull-wheel and pulley attached to the derrick play an important part.

The top of the chamber, when adjusted in the well, usually extends about six or eight feet above the derrick floor. Sometimes a tub or a half barrel is placed on the top, with a tube running from near the top of the tub into the oil tank. Sometimes there is simply a tube connecting with the chamber, at an angle sufficient to conduct the oil and water into the tank.

The testing of an oil well is often very tedious and perplexing. It may yield nothing but water, and that at such a rate as almost to induce the workmen to believe there is an underground communication with the river. Perhaps there is a vein of water below the seed-bag, or it may be there is a crevice or fracture in the rock just at the point where the seed-bag is placed, permitting the water to run in from above. The sucker-rods are drawn out, then the chamber, piece by piece, and the seed-bag adjusted to a new place, and the whole put in again. The second trial, perhaps, succeeds no better than the first. New theories are advanced, and new trials made of re-adjusting the tubing. Perhaps, in their desperation, the workmen may put the pump in operation, and continue for weeks pumping salt water, when at last the oil will begin to flow, rewarding their persistent efforts with a copious yield. The philosophy in this case seems to be, that the cavities containing the oil are a long distance away, and that the avenues leading to them are filled with water in a sluggish state, with, perhaps, a stratum of gas preventing a single drop or globule of oil finding its way through the water. When this latter is removed by the pump, the vacuum opens the way for the oil, and if the cavity is sufficient, the flow will be copious.

Often, however, this persistent pumping proves fruitless, and resort is had to re-adjusting the tubing and so on for weeks and months. Instances are on record where this re-adjustment has taken place as many as fifty times, resulting in success at last. But the result is oftentimes failure after all the labor and experimenting that can be brought to bear. Either there is no oil, or the manner of tubing has been defective throughout, and the adventure is a total loss.

In the shallower wells the proportion of water thrown out by the pump is very great. Veins of water are sometimes passed below those that yield oil. Sometimes this bit of experience proves fatal to many a promising oil vein. In this case, attempts are sometimes made to stop up the well below the vein of oil, but this is difficult, as the vacuum above assists the upward pressure of the water below, and the obstruction yields to the combined attack and gives way, deluging the well with water again. In the deeper wells, especially those reaching into the third sand rock, particularly where the seed-bag is attached within the third sand rock, this difficulty is not so much in the way. In fact, it is the opinion of many operators of experience and judgment, that there is no water found in and below the third sand rock, and that if the well be properly tubed, the pump will present no water at all, but a constant stream of pure oil. Let the value of this theory be what it may, the fact is on record, that from deep wells water is often entirely excluded.

Another difficulty that meets the workmen occasionally is, that without any known cause the well will cease its yield, where the tubing has not been disturbed. Sometimes, doubtless, this result takes place because the oil has all been exhausted from a small cavity, or crevice of limited extent. But this is not always the cause. It may result from the settling down of the sand and mud from the sides of the well, or from crevices in the rock over the vein, clogging it up. In this case the remedy would consist in withdrawing the chamber, and, with the sand-pump, extracting the sediment, and perhaps boring the well somewhat deeper. Instances are observed, too, where wells that had yielded plentifully have gradually lessened their supply, and finally ceased altogether after having thrown out for a time minute particles of parafine, thereby suggesting the idea that this substance had gradually accumulated until it had closed the orifices. In such cases two remedies have been sugested, and either one has in a few instances resulted in success.

The first is by the use of steam. A small gas tube is put down the chamber, and a jet of steam is forced down from the engine, until the gradual heat is supposed to melt the parafine, and thus open the veins. Under favorable circumstances this is practicable. Parafine melts at 112°, and, if the water is pretty thoroughly exhausted, that amount of heat might be attained by the injection of steam for that length of time.

The other remedy attempted has been a kind of torpedo, or strong water-tight box filled with powder, and exploded by a galvanic battery. This explosion is supposed to distend the opening, enlarging the veins, and preparing for the flow of the oil to the region of the pump.

The simple fact that these remedies often fail is no evidence against the soundness of the philosophy on which they are based. The idea of firing upon a beleaguered city or garrison by means of mortar batteries is certainly feasible and philosophical, yet it has its difficulties and discouragements. Perhaps not one-half nor one-fourth of the shot reach their destination, but fall harmlessly in out of the way places, and no one knows what becomes of them. Still mortar fleets and batteries will be used under certain circumstances, and with a reasonable prospect of doing good service. And it is by trying various kind of experiments, that seem calculated to answer the end in view, that difficulties, like those spoken of, can reasonably be hoped to be removed.

The question here becomes a very interesting one whether in pumping, wells ever interfere with each other. At the first stages of proceedings it was thought that they did not, and that all fears in regard to this were groundless. But subsequent developments indicate that there is a connection in some instances by means of which oil is drawn from one well to another. In one instance, the vein or cavity that had been drawn from, for a length of time, by a successful borer, was tapped by an enterprising neighbor, when all proceedings were at once stopped on the part of the first, the second adventurer taking all the oil. The result was that a compromise was affected, the two wells pumping alternate days.

This state of affairs has given rise to a number of curious questions in casuistry, or "cases of conscience," as the old fathers called them, not laid down in the books, and that have been generally adjudicated without recourse to courts of law. So far there seems to be a disposition to do justly, to deal equitably with each other, and to consider the business a common one, in which they are all interested. It is something of the feeling that binds people together in new countries, where they feel that mutual concessions and looking after the general welfare is the duty of each individual.

One of these cases arose in this wise. A well had been pumping with considerable success for a length of time. The body of water had been heavy, yet by vigorous pumping it was kept down, and a fair proportion of oil was obtained. Another party bored a well in the neighborhood, and on commencing to pump found nothing but water. He persevered for a time although the result was not changed, yet to his neighbor, the first operator, there was a decided advantage as the quantity of oil was very materially increased. The second party finally concluded to abandon the work as unproductive.

The first then proposed to pay the second for pumping water, as the advantage to him was very great. This arrangement was entered into, when, after a time, well number two commenced yielding oil, and number one began to fail, and its proprietor discovered that he was paying his neighbor for robbing him. And yet it was involuntary robbery brought about by the mysterious operations of the rocky labyrinth beneath.

As to the regularity of pumping there is a difference in wells, particularly among the smaller ones. Generally it is of the greatest importance to pump continuously, both day and night, in order, at times, to keep down the volume of water, and at other times because the amount of oil gradually increases day after day, by means of such pumping. In the earlier stages of the business, and when wells were not bored very deep, this was particularly observable. If they "rested the Sabbath, according to the commandment," on Monday morning the proportion of oil was small, but each day of the week it increased until Saturday night, indicating that there was a connection with water veins somewhere, and that steady and severe pumping was necessary to overcome the water.

But there are other instances where no damage has resulted from a cessation of pumping for a time; and even where the pump has rested altogether during the night, experience proving that as much oil could be pumped in twelve hours as in twenty-four. This state of affairs is, no doubt due to the fact that the water was excluded in large volume, and that the oil was contained in a cavity of limited extent, fed by small fissures, and that when this cavity was nearly exhausted by a day's pumping the rest of the night permitted the cavity to fill up and be ready for the next day's operations. This condition of things belongs rather to wells of small capacity; in the deeper and large producing ones, stopping operations for any great length of time has sometimes operated disastrously. Instances are known where a rest of a few weeks or months has entirely ruined the well, or, as the workmen term it, the oil has been lost. Perhaps this result has occurred generally where numerous other wells have been pumping in the same neighborhood, that have attracted the oil in their direction, until it has ceased to flow in the old channels. In this case, either the inducement is greater in the new, or the old veins have become obstructed from mechanical causes.

In all these matters observation and experience are the best teachers, and the true policy is to operate with a given well in such a way as close observation shall indicate to be best under the circumstances of its location and surroundings.

Another way of bringing the oil to the surface, where sufficient gas is present to answer the purpose, is to remove the ordinary chamber, and in its stead put down a gas pipe of a half or three-quarters of an inch in diameter, with the usual seed bag attachment. This small tube so confines the gas that often a small flowing well is extemporized that will yield as much oil daily as could be pumped in the same time, and with little or no expense.

A late invention has been patented that does away with the pump as a means of raising the oil to the surface, by using an air-pump in forcing condensed air into the well, and thus acting in the same manner as gas in forcing the oil up the pipe. The arrangement is simple: two pipes are put down into the well parallel to each other. One, called the discharge pipe, is an inch and a half in diameter, having an enlarged cup at the lower end to receive the oil. The other, called the blast pipe, is connected with the air-pump above, whilst below it is turned up like the short leg of a syphon, so as to enter the cup-like opening of the discharge pipe, and tapers at the nozzle to one-third of an inch in diameter. This arrangement is designed particularly for wells of great depth. In some instances where it has been tried in wells that have been obstructed it has proved a valuable aid in overcoming the obstructions, and preparing the way for the ordinary lifting pump. It has not been sufficiently tested, however, to enable the public to decide upon its final merits. It is claimed for this arrangement that it acts not only "by the momentum due to its velocity; but, mingling with the oil, lessens the weight of the ascending column, and thus aids the work."

A difficulty has arisen in regard to the quality of the oil produced in some wells. Sometimes they will pump or flow what is termed "muddy oil," or "riley oil," the latter term being more provincial than classical. In this condition of the oil it is, as the term would indicate, mixed with mud—not only sand, but an argillaceous, adhesive mud, that is held in precipitation, though not in solution, by it. So firm and tenacious is the grasp of the oil in this mischievous compound, that it is extremely difficult to separate them. This state of affairs is excessively annoying to oil men. The matter of separating the mud from the oil is difficult and perplexing, and unless it is perfectly accomplished the oil is damaged in the market. It is the opinion of some that this is produced by muddy water running into wells that have been abandoned, and finding its way to the oil cavities, but it seems due rather to the influence of the "mud veins" referred to in chapter ix. The agitation of the gas, and oil, and water, either in pumping or flowing wells, may, under certain circumstances, dissolve and set free the substance of the mud vein, and thus produce the phenomena of muddy oil.

The cost of tubing a well six hundred feet in depth, and inserting pumping apparatus, may be stated as follows:—

| | |
|---|---|
| Six hundred feet of chamber at 60 cents | $360 00 |
| Pump barrel | 35 00 |
| Six hundred feet of sucker-rods, at 17 cents | 102 00 |
| Two pairs of gas-tongs | 12 00 |
| Total | $509 00 |

In connection with pumping, the preparation of tanks or vessels to receive the oil as it comes from the well is one of great importance. Generally the oil is mingled with a large proportion of water, which must be separated from it before it can be barrelled and sent to market. For small wells, this tank is usually constructed of planks, keyed together within a frame work, and made water-tight. Into this tank the oil and water flow from the discharge-pipe of the well, where the superior gravity of the salt water causing it to sink to the bottom, the oil is found at the surface. The water is then drawn from the bottom of the tank, and the oil from above is received into barrels.

This arrangement answers very well in connection with wells of small capacity, and when the demand for the product was brisk, but in process of time, when large wells were opened, and particularly flowing wells, the matter of tanks became a very important one. A plank box no longer answered the purpose; it might be filled in an hour. The plan of hydraulic cisterns under ground, like rain-water cisterns, was suggested, and experiments made by workers in hydraulic lime, but the experiments failed; the petroleum is of so subtle and permeating a nature that it finds its way through substances that will retain water with ease.

An attempt was made to construct large tanks of upright hoops banded with iron, but having earthen bottoms puddled with potter's clay, and with a stratum of water on the top of the clay. But the oil men soon found in the gutters and inequalities around the cistern, oil to such an extent that a profitable trade might have been carried on in gathering it up. This plan was abandoned. The general plan now, is to construct an immense tub of planks, hooped with iron, having a plank bottom. These tanks are sometimes twenty-four feet in diameter and sixteen feet high, and will contain, perhaps, ten to twelve thousand barrels of petroleum. Oftentimes a number of large tanks are found connected with the flowing wells; usually, however, they are not so large, but contain, perhaps, three or four thousand barrels, and increase in number what they lack in quantity. The latter plan, although a little more expensive than the former, is the safest, as an accident happening to one may not be communicated to the others, whereas, if all the product of a well is stored in one tank a slight accident may destroy it all.

In these tanks the petroleum is received from the wells, separated from the water, and retained until it can be put into barrels and sent to market. A large amount of tankage at the wells renders the market more steady and regular, as it can be kept on hand, and not forced on the market for want of a place of storage. This, together with the amount of capital now thrown into the oil trade, has latterly prevented the panics in the market that at one time characterized it, and operated so disastrously to many persons engaged in the business.

Sometimes, in very cold weather, the separation of oil and water becomes slow and difficult, and to remedy this the steam-pipe of the engine is conducted into the tank, and by the assistance of a drum at the bottom, the heat necessary to promote this separation is obtained.

The fuel used for the engines is bituminous coal; but this is expensive, and resort is had to the use of the gas that escapes from the well. In some wells this is amply sufficient for the purpose, in others it answers in part, the lack being supplemented by coal. The apparatus for collecting the gas is very simple. The oil and gas together, as they come from the well, are received in the top of a barrel or hogshead. A pipe from near the bottom conveys the oil to the tank, and another from near the top conducts the gas to the fire-chamber in the engine. The pipes are so regulated that about the proper amount of space is preserved at the top of the barrel to receive the gas, the pressure of the oil forcing it to the fire-chamber. The only danger to be guarded against is fire, but, as yet, this has not proved a formidable danger, and the plan works to very great advantage.

# CHAPTER XI.

## FLOWING WELLS.

THE business of boring and pumping was proceeding regularly and encouragingly, when a new feature presented itself that was most startling, and, in some respects, disastrous. Wells were being bored along Oil creek, the Allegheny, French creek, and their tributaries, and were in all stages of development. Some had simply erected their derricks, some were an hundred feet in the rock, and others actually pumping, when the change came that wrought ruin to the hopes of many an ardent operator.

In the Oil creek region, some of the smaller wells having been exhausted, resort was had to deeper boring. One hopeful theorist imagined that if the desirable fluid came from a very great depth, it might be good policy to seek it in a stratum still nearer its rocky home. So down he penetrated, regardless of the "fine show of oil" that presented itself by the way, until, at the depth of five hundred feet in the rock, a vein of mingled oil and water was reached that literally forced the boring implements from the well. The oil continued to flow with a constant stream, after this sudden exodus of the implements, rising to a height of sixty feet above the surface of the ground, and was occasionally accompanied by a roaring sound, like the Geysers of Iceland.

Here was a new feature in oil operations. Heretofore the production had been the result of slow and painful pumping, and this at the rate of a few barrels per day; here was a spontaneous yield of hundreds of barrels daily, without expense or labor. Was it possible to continue this inexpensive process? Was it possible to carry on the trade by simply perforating the rocky crust, and then trusting to the mysterious forces of nature to pour forth the petroleum in indefinite quantities? It seemed so from this exhibition, and yet, if it were so, a complete revolution must take place in the business. Oil would become cheap as water, and the business become at once unprofitable. Notwithstanding all this, the idea was a brilliant one, and men seized upon it with avidity. The idea of flowing wells for the spontaneous production of petroleum, once inaugurated, must be pursued at once, and with persistent energy. There was not only a spontaneous yield, but a yield in enormous quantities. So a "pumping well," as it was called, was voted a slow institution, and parties who had been satisfied with the old order of things, and were growing rich on the proceeds of pumping, renewed the operation of boring near their old sites, and many, at the depth of the first flowing well, met with like success. Parties also, that were boring, continued on in spite of all oil indications, to the depth of five hundred feet and beyond, and many of them were rewarded by opening the way to veins of oil and gas that gushed forth spontaneously and continuously. Every man on the creek was anxious to have a flowing well, although the product might remain useless upon his hands. The dark green fluid represented wealth; it had made many rich, and large quantities were desirable in any event.

But the operation of these wells was disastrous to the trade generally during the first six months of their flow. Their enormous yield had the effect of bringing down the price of petroleum to so low a figure that pumping wells were at once closed. The proceeds would not pay for the fuel—scarcely even for the payment of the workmen, and at once the entire business, with the exception of a few wells at Franklin and French creek, that yielded heavy lubricating oil, was confined to the valley of Oil creek, then the region of flowing wells. Oil at once fell to a price much below that of the barrel that contained it. Parties sending their barrels, or "packages," as the oil men call them, to these flowing wells could have them filled at one cent per gallon, or less if they had the heart to ask it. It could be bought for less at the wells than common creek water has latterly been selling at in the streets of Franklin.

Of course, there was at once a panic in the oil region. Whereunto would this matter grow? Some wise heads predicted that in a few days the stream would cease to flow, and sat down to watch its decreasing volume, growing "small by degrees, and beautifully less." But this was of no avail. The stream seemed to increase rather than decrease. It was like the countryman sitting upon a dry-goods box on Broadway waiting until the crowd got by. Others, equally wise, thought they saw in this new phase of things not only disaster, but utter and total ruin to the oil business. The market would be overstocked; it would not be worth an ambitious man's attention; the ship was sinking, and must be abandoned at once. One extreme generally follows another, and persons who had a few weeks before considered the oil business the great, hopeful business of the, age now condemned it as the prince of humbugs, and Venango county the great "Vanity Fair" of the earth,

"So swift trod sorrow on the heels of joy!"

On the Allegheny and French creek, wells were abandoned, leases forfeited, and machinery removed as the operators shook the dust from their feet, as a testimony against the wicked and deceptive region where they had spent their money for nought. There was some ground for this discouragement, still it was not altogether justifiable. It would have been better to take in the sails, bolt down the hatches, and wait until the storm had passed by; or at least attempted to weather it out.

These flowing wells were found at the depth of about five hundred feet in the rock, and usually in what is termed the "third sand rock." They are tubed as other wells, with the exception that the pump attachment is wanting, and the tube has an elbow some eight feet from the derrick floor, from which the discharge pipe leads into the tank. In some of these wells the stream is pure oil, in others mingled oil and water, in each instance accompanied, of course, by gas, that is the motive power. The quality of the stream, whether pure oil or oil and water, depends, as will be hereafter seen, on the particular point in the vein perforated, in relation to the cavity containing the oil.

The philosophy of flowing wells may be readily understood from a diagram. We may imagine cavities in the rock beneath of every conceivable form and size. The strata that is generally termed the oil bearing rock may have been contorted and broken by some internal convulsions of nature, through the agency of heat and gas, so as to form caverns of great extent. These caverns have connected with them seams, and leaders, and cavities usually termed veins. In these cavities the oil is collected and stored away. But water also forces its way through crevices, and is found in connection with the oil; gas also will be found in the same receptacles. Now, it is not only probable but absolutely certain that these three fluids will be found in the cavities of the rock beneath, in the same relative position that they are found in the covered tanks around the wells. The water will be found on the bottom, the oil next, and gas on the top, pressing against the top or roof of the cavity, and becoming greatly compressed by the forcing in of the water below. Let the engraving then repre-

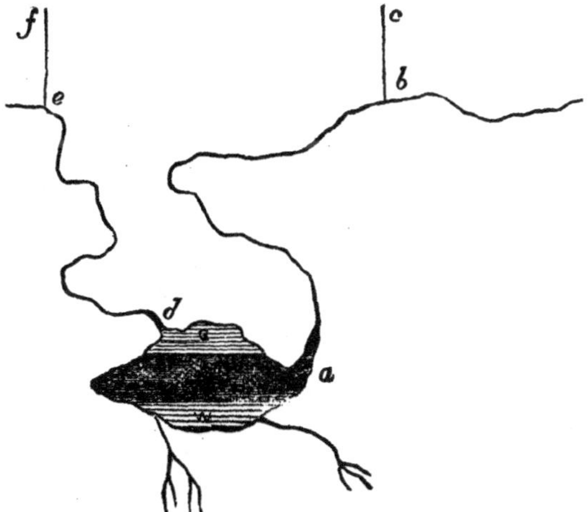


sent a section of the rock strata from the surface downwards. The cavity, of course, is imaginary, yet conceivable, containing water below, then oil, with gas in a compressed state above. If we conceive a small vein or cavity leading in the direction from *a* to *b*, filled with water, with a well sunk (*c*) from above, and striking this vein at *b*, we will see at once that a flowing well will be the result. The tremendous pressure of the confined gas at the top of the cavity will force the oil and water downwards, and through the vein to the point where it is tapped, and so up through the chamber with a force proportioned to the volume of the gas and the constancy of its supply from other cavities; it may be far down in the regions below. If the cavity indicated be very large, and the connection with other cavities be by means of large crevices and veins, we may suppose that the supply will be constant, and the flowing well continue in operation for quite a length of time, though it can hardly be expected to be permanent.

If the vein tapped connects with the cavity in the region filled with gas, or at the point *a*, there will be an explosion of gas, and it is barely possible that the reaction of the gas on the oil may cause it to flow, particularly after the water has been removed from the vein with a pump; but this is doubtful, and especially is it questionable whether, under such circumstances, the oil would flow for any length of time. Still tapping a vein like this would readily beget a pumping well, for after the gas was removed, the pressure of water from other veins would raise the oil to the sphere of the influence exerted by the pump, and there would be a pumping well accompanied at times with considerable annoyance from gas. And this is not an unusual result. Sometimes the pump works with difficulty on account of the presence of gas, and occasionally by the introduction of a very small tube, oil has been made to flow in small quantities.

The history and amount of production of some of these flowing wells is interesting. The first one was developed on the McElhaney farm, and is usually known as the Funk well, being bored by a man of that name. It was opened in June, 1861, and at once commenced flowing at the rate of two hundred and fifty barrels per day. In the face of all past experience, and all ill-boding prophecies of speedy failure, it continued to yield a constant supply for one year and three months. In the fall of the same year a flowing well was opened on the Tarr farm that completely eclipsed the Funk well. This was the Philips' well, and commenced at two thousand barrels per day. This was followed by the Empire well at the modest rate of three thousand barrels daily! This was the "eagle's highest flight," and the wells stayed at this in quantity though not in number.

These were followed in irregular succession by others. The Noble well yielded twenty-five hundred; Caldwell, eight hundred; Maple Shade, one thousand; Jersey, five hundred; Coquette, fifteen hundred; Reid, one thousand —this latter is on Cherry run; and so on down to ten barrels per day, literally fullfilling the words of Job, "The rock poured me out rivers of oil."

Some of these wells were bored under extreme difficulty, and sore discouragement, arising from want of means and straitened circumstances. It must have been a glad sight to see the mighty current gushing forth after so much labor, and toil, and discouragement.

The history of the Funk well is briefly this. Captain Funk was the possessor of a few modest acres on Oil creek; but lived in Titusville a few miles above. When the oil business commenced, encouraged by the success of some of his neighbors, he resolved to attempt putting a well down on his own farm. He gave a lease to a man by the name of McElhaney who undertook the task of drilling a well by the tedious process of working the spring-pole by his foot. This plan worked well enough until the well was perhaps one hundred feet deep, when the labor became too burdensome. A horse-power was substituted, and the work carried forward to completion.

The Sherman well, also, has an interesting history. Mr. Sherman came to the oil region a man of very limited means, but could command a small capital belonging to his wife. He obtained a lease on the Foster farm, and commenced to put down a well by hand. He had a fine show of oil, but long before reaching any considerable vein, the wife's funds gave out. Working by hand was no longer possible on account of the depth of the well, and there was no means at hand of purchasing a horse. After working until something should turn up, an interest in the well was finally exchanged for a horse, and the work proceeded. But after a while the work became too onerous for the poor horse, and the well was once more at a dead stand. At last a farther interest in the well was bartered to two men who owned a small engine, and the work once again went forward. But coal was expensive, and none of the partners were able to purchase, and the work was suspended. A sixteenth interest was now offered for sale in order to purchase fuel, when, after waiting for a time, it was disposed of for eighty dollars and an old shot gun. The money and perhaps the shot-gun were just about expended, and the spirits of the partners down to the lowest ebb, when the bit plumped into a cavity that at once yielded fifteen hundred barrels per day. The partners in faith and sorrow, now had no longer need for either the dilapidated horse nor the little engine that had done them such good service; nature worked the pump on her own account, and all the owners had to do was to barrel the oil, and receive the proceeds. This well decreased to seven hundred barrels, and at the end of about two years became a pumping well. It is now a valuable well of that description, and is doing its owners good service.

These wells were unequally distributed along the creek. The Maple Shade, Jersey, Keystone, and Coquette are all on the Egbert farm; the Sherman on the Foster, and the Noble on the Ferrel farm.

Flowing wells were at the outset necessarily accompanied by a great loss of oil. At first wells were bored with the hope but not with the certainty of oil, and the tank was usually a secondary consideration. When the first wells were opened, producing such grand results, there was little or no tankage ready to receive it, and the oil ran into the creek and flooded the land around the wells until it lay in small ponds. Pits were dug in the ground to receive it, and dams constructed to secure it, yet withal the loss was very great. In addition, oil fell to so low a figure that few comparatively were willing to hazard purchasing at any price. The consequence was that the river was flooded with oil, and hundreds of barrels were gathered from the surface as low down as Franklin, and prepared as lubricating oil. Even below this point oil could be gathered in the eddies and still water along the shore, and was distinctly perceptible as far down as Pittsburgh, one hundred and forty miles below.

These erratic institutions were almost as difficult to manage as was Pegasus of old. They "played fantastic tricks" when least expected, throwing the oil over the workmen, deluging the region around, and, in one case, where the vein of oil was suddenly opened, setting fire to the machinery, destroying the workmen's tools, and raging with appalling fury.

One notable instance of this kind must here be related. It is a tragical chapter in the history of petroleum operations, the result of which settled like a gloomy incubus upon the oil regions at the time. It was at the first flowing well that was struck, and was known as the "Burning Well," and located on the "John Buchanan Reserve," a part of the "Buchanan farm." The well had reached a depth of over three hundred feet when a column of gas rushed up that seemed to fill the whole region around. This gas was ignited by the fire of the engine that was pumping the Wadsworth well, eighty to one hundred feet distant. The first explosion was from the tank at the Wadsworth well, containing about one hundred and fifty barrels of oil. Immediately afterwards "the well exploded," as expressed by a bystander, with a shock like that of an earthquake, shaking every house on the Buchanan farm. All this occurred within thirty minutes after striking the vein in the well; in the meantime the mingled oil and gas was pouring from the orifice with terrible fury. It seemed as though the earth was vomiting flame threatening to fill the whole valley as with a sea of fire. The fiery column reached far above the derrick that was soon consumed, accompanied with dense volumes of black smoke, roaring like a hurricane, turning and bending in every direction as the wind veered from one point to another.

There were at the time of the explosion from ninety to one hundred persons standing around the well, many of whom were soon enveloped in the flames and saturated with the scalding burning fluid. Many of these persons had their clothing saturated in an instant with the spouting oil when the flames at once seized hold of them, and rendered them helpless in its fiery grasp. Of thirty-eight men, more or less severely burned, eighteen died. Among the latter was Mr. H. R. Rouse, one of the proprietors of the well, and one of the most enterprising and energetic of the business men along the valley of Oil creek. Some of the victims could not be removed, and were left in the midst of the flames until well nigh consumed.

The scene was grand and awful, and had it not been for the suffering and loss of life that attended it, would have been interesting beyond description. Now the flame and smoke seemed to play upon the summit of the lofty hills and over the tree-tops, and soon they would swoop down into the valley like an eagle on his prey, accompanied by the hissing, roaring sound that added to the terrible accompaniments of the scene.

This fire continued five days, and was finally extinguished by digging up earth and carrying it upon blankets, and smothering the flames. After the fire was extinguished about sixty feet of chamber were put in the well, when it flowed about twenty thousand barrels, and ceased. After this a new chamber was extended down to the second sank rock, about three hundred feet from the surface, and a pump inserted, but no oil was found.

Subsequently it was bored to the depth of five hundred feet, but as yet without success.

Other cases of fire have occurred on the creek resulting in much damage, but without the loss of life, save a few instances, where oil had been used as a kindling material in houses. A remarkable case of fire occurred at Oil City, where numerous boats taking fire, floated down the river, and consumed the bridge over the Allegheny river at Franklin.

With all the disadvantages to smaller wells, and the loss of oil from irregular operations, these flowing wells have done a good work, for the general interests of the trade. Before their advent in the field of operation, the process of introducing oil generally throughout our own country was slow and irregular. The people had doubts in regard to the permanence of the supply. The price was too high almost to justify doubtful experiments, and the old tallow candle was battling manfully for his ancient rights, and for the defence of his tottering throne. The exports were very small indeed, arising from the same cause; and it needed some mighty impulse to introduce the oil to general use both at home and abroad. It required some greater stimulus, too, than the pumping wells to bring petroleum into use for other purposes besides that of a light producer.

The reign of flowing wells was just what was required. They deluged the popular mind; they deluged the country. Prices came down at once so that the article, beautifully refined, could be purchased for twenty cents per gallon, and burned freely in the humblest households. The tallow candle was no longer king. With the ruin of his empire he came down from his throne gracefully, and "retired to some sequestered spot" to come no more into public view. Petroleum was at once introduced to all parts of the country, and experiments made upon it for various purposes, causing an increased demand to be made for it from all classes of the community. It was introduced into foreign countries, and its value and importance tested there, until it became a necessity there as well as at home.

The result has been that petroleum came to be considered a necessity, and the demand became regular, or rather continued to increase even after prices were raised to the old rates that prevailed previous to the reign of flowing wells. The people could not go back to the old days of candles and lard oil lamps, after having experience in a substitute, so much superior. It would be like going back from civilization to barbarism. And thus the trade was at once firmly established. At one mighty bound it took its place as one of the great, if not the greatest, staples in the world's market, and its success was placed beyond a peradventure.

But the question naturally arises in regard to the flowing wells, "Will they not fail?" "Can they be looked upon as permanent?" Most undoubtedly they will all ultimately fail. It cannot be hoped that they will continue their spontaneous yield for any great length of time, for, from the very principles on which they act, they must soon exhaust themselves. True, artesian wells flow on from year to year, without exhaustion. But they act on different principles. The water from them is forced up by the pressure of a fountain head that simply forces the water to its own level, and as long as the fountain continues the well must flow. But flowing wells appear to be due to the simple pressure of gas confined and compressed with the oil, and as soon as this is exhausted the well must cease to flow.

This is not only probable in theory, but has proved correct in fact. The history of all flowing wells, so far as they have continued long enough to have a history, has been, that they have gradually declined in the yield, until they have subsided altogether. And this will no doubt be the history of every flowing well that will be opened. Nor is this an undesirable feature, if we take in the interests of the entire oil region at a glance. The few who have an interest in these large concerns may desire their continuance, yet it would be at the sacrifice of smaller or pumping wells; and a good pumping well should satisfy the desires of any reasonable man. The whole business would be more equally adjusted, and the price of oil be steadier and more regular, if all the wells in the oil valleys were operated by pumps. Still these flowing concerns are yet doing good service at times, breaking forth and attracting attention to new territory, and developing regions where little attention had hitherto been directed.

We need not necessarily suppose that because a well ceases to flow the supply has been exhausted. The quantity may have been reduced, and the gas withdrawn to such a degree that it has no longer sufficient expulsive power to force the oil to the surface. In this case a pump-barrel and sucker-rods inserted, and the assistance of the engine invoked, it may yet prove to be a good pumping well. It may do good service still. And history corroborates this theory. Almost all the wells enumerated, after indulging in ground and lofty tumbling for a time, have settled down into regular habits. They have sown their wild oats and become regular and orderly in their habits, and bid fair to live to a ripe and unctuous old age, bearing joy and gladness to those whom they serve.

It is probable, however, that when these wells come down to the degree of pumping institutions, they will not yield as largely as many that have not aspired to the dignity of flowing wells. The supply, it is likely, was drawn in the days of their glory from enormous cavities that have been chiefly drained, yet still are fed from other cavities beneath, by means of small veins or leaders in the rock; and the pump can only withdraw this new supply as it is gradually presented by these small apertures. A well, then, that has yielded in its palmy days two or three thousand barrels daily, and comes down to ten or fifteen barrels, and this extracted through the persuasive power of a pump, may continue to yield, unless its crevices become filled up by mechanical means, for many years, and prove as valuable as those that have been pumping wells from the beginning.

Attempts have been made at various times to close the chambers of flowing wells by means of a stop-cock, and thus draw upon their ample resources just as the supply was needed, but the attempt did not succeed well. The effect was found to be damaging to the well. In some cases the supply began to fail and threaten total destruction to the value of the well. In such cases, it is probable there were connections with other cavities and veins that had been pierced by neighboring borers, and that through these cavities and the influence of neighboring wells, the supply was diverted in a different direction, and the old channel completely abandoned. Under such circumstances, the application of a pump, and continuous pumping for a length of time, might recall the truant oil to its former channel. Ordinarily now the workmen prefer letting the oil go to waste if the flow precedes the erection of tanks, until they can provide means for securing it, rather than run the risk of closing the well. Latterly, however, the prudent borer has his tanks prepared by the time his well is ready to flow.

Usually the flow from these wells is steady and regular, although pouring out with terrible fury and energy. It seems, at times, as though the power was sufficient to turn a small mill. But withal it is regular and constant, yielding for weeks almost the same quantity from day to day, and from hour to hour. Sometimes this yield is pure oil, scarcely a trace of water being present. This probably depends somewhat upon the correctness and security of the tubing, and somewhat upon the nature of the rock, as regards the presence of openings and crevices through which water may be enabled to percolate. This seems to be the result particularly in deeper wells, and those yielding large quantities of oil. It may depend somewhat upon the size of the cavity in which the oil is found, and the manner in which the vein that is struck connects with this cavity. If the connection is with that part containing oil, and above the water, the result would be pure oil. Often, however, the oil is largely mingled with dense salt water, that must be separated in the tank before the oil can be put in packages for the market.

Sometimes there is a kind of intermitting flowing well. The oil and gas will flow for a little time, then cease altogether for about the same length of time, and then flow forth as before, presenting all the phenomena of an intermitting spring. One in particular may be specified for the regularity and system of its operations. It will remain quiescent for about fifteen minutes, when there would be heard the sound as of fearful agitation far down in its depths. This rumbling and strife would then appear to approach the surface for a few moments, when the petroleum would rush from the orifice, mingled with gas and foam, almost with the fury of a round shot from a rifled cannon. This furious flow would continue for fifteen or twenty minutes, when it would suddenly subside, and all would be peace again. This alternate rest and motion would continue with great regularity day and night, yielding, perhaps, one hundred and fifty barrels per day. In other instances there will be interruptions of days, and even weeks, when the flow will be continued as before.

In these cases it is possible that there is a peculiar conformation of the veins with relation to the cavities containing the oil, by which the supply of oil and gas is exhausted for a space of time, when the well remains quiet until the cavity is filled up again from other sources.

Let the cavity be supposed to contain the three fluids as before, but connected with a second cavity by a crevice at *c*. If the well, *a*, pierce the vein, *d*, it will flow until the cavity is nearly emptied, when it will cease until the first cavity is refilled from the second. See page 158.

Perhaps the most singular phenomenon connected with flowing wells was what was called the "Sunday well," from the fact that it flowed constantly during the week days, but ceased on the Sabbath. It was regarded with some little superstition by the workmen engaged at other wells, but soon ceased to attract much attention, further than being regarded as a great curiosity by some, and by others as involving a curious and interesting philosophical question. No doubt there is a very general connection existing between the cavities containing oil, by means of veins and small seams running in every direction, that affect in a greater or less degree the operations of pumping and flowing. Sometimes this

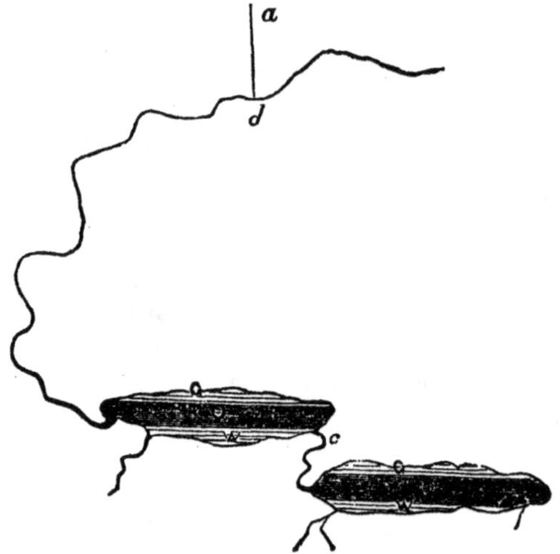


influence is so slight as to be imperceptible; at others, it is felt in different ways. In the case of the Sunday well, the agitation of pumping, and the withdrawal of water from particular veins, induced such a condition or equilibrium of oil and gas in the cavity underneath, as to produce the phenomenon of flowing during the time of pumping. On Sabbath the pumping ceases, the agitation is no longer moving, and stirring, and disturbing the labyrinthine veins in the rock, the water accumulates and the flow ceases. On Monday morning the agitation commences, the water is withdrawn, and the flowing commences as before; so that, after all, the mysterious operation of this well was but the result of counterbalancing forces operating in the cavities beneath.

Another thought in relation to flowing wells. The strong probabilities are, that as the number of wells bored in any given locality increases, the number of flowing wells will decrease, until they cease altogether. If they are due to the exhaustive force of the gas imprisoned beneath, and there is no room to doubt this, then as the number of wells increases the gas will become gradually exhausted, or will find new avenues of escape, not only from pumping wells, but in wells that are absolute failures, until the gas force becomes like the bow that has been too long held in tension, its power will be lost, and pumping wells must be the final dependence. This is becoming the case in many places already, and will, no doubt, be the case finally throughout the oil region.

---

## CHAPTER XII.

### MEANS OF TRANSPORTATION.

As the business began to increase, and its magnitude to develop, the question of transportation became a very important one. Although not bulky in proportion to its value, it has many unpleasant features connected with it. Being generally carried in wooden barrels, or "packages," to use the modern term, its penetrating qualities render these barrels unpleasant and disagreeable in the handling and porterage. And as this leakage is accompanied by the escape of gas, the transportation is attended with considerable danger from fire, involving loss of freight, and in the case of steamboats of life itself. Still so far no very serious losses have been incurred, although special care has been exercised by those engaged in the business. And the annoyance of working among greasy barrels, even, seems to be very readily overcome.

Petroleum appears to have been transported at first in small bottles, for a very small portion answered the demand. In almost every house a broken pitcher or teapot was filled with it and was replenished when necessary. Mr. Cary was a bold man in those days to cut loose from the old habits of caution and fill two five gallon kegs for transportation to Pittsburgh. But bold spirits usually open the way for new enterprises. Could this early dealer in oil but know of the thousands of barrels that now find their way to the same market, sometimes in a single week, it would almost cause his old bones to rattle in their quiet slumber, near the scene of his early enterprise.

When the business began to develop, the Allegheny river was the only way of conveying the product to market. The distance to Pittsburgh is about one hundred and forty miles. Much of the oil was refined there, but still more was shipped to Philadelphia, New York, and other points. In times of high water steamboats run up to Oil City, and at the same time other species of craft run down. Transportation by the river is perhaps the cheapest of all modes of reaching the market; but in this fast age is a little slow. The steamboats used are of the kind usually termed stern wheels, that is propelled by a single wheel which is rigged at the stern of the vessel. The petroleum is stored in the hold and on the lower deck. Some of these boats carry five hundred barrels. The cost of transportation varies according to the number of boats running and the demand for carriage; as in the oil region men are not usually slow in taking advantage of circumstances. The price ranges from forty cents to one dollar per barrel.

Another mode of transportation is by loading the oil upon rude flat boats, resembling coal boats, or the old "broad horns" of the early settlers, and towing them down the river with a steam-tug. Sometimes these flats are loaded with bituminous coal, far down the Allegheny and towed up; thus realizing a profit both on the upward and return voyage, as the same flat answers for both purposes. These tugs are a very slow but powerful species of craft, small in outline but propelled by powerful engines, that in their own peculiar way enable them to perform good service in this kind of business. One of these tugs will sometimes convey to market three thousand barrels of oil.

The flat-boats have always been an institution on this river and were not slow to avail themselves of the new business that has fallen in their way. They are often towed by horses up the current, and allowed to float down with the current. Horses are attached by very long lines, and sometimes walk on the shore, and at other times in the stream itself. Whenever the water is deep, and it is desirable to change the motive power to the opposite side of the stream, the horses are taken on board and transported to the other side. Sometimes these flat-boats are pushed up the stream by the boatmen themselves; or, as they term it, "make progress moving backwards." The boatman with a long pole set in the bottom of the stream, and with the other end braced firmly against his shoulder, commences at the bow of the boat and walks slowly towards the stern, pushing as he goes. He then walks back and repeats the operation. Several men being engaged in this work, on each side of the boat, the unwieldy craft moves slowly and regularly forward.

There is another species of craft that is pressed into the service, which is more primitive in its construction. It is formed of two gunnels hewn or sawed from a log, with the ends cut so as to turn upwards like sled-runners. These are connected together by scantling and a bottom spiked on. Studding is then set up along the sides or gunnels, and rough boards spiked to these, and after being rudely caulked the boat is ready for use. Oil barrels are rolled in, when the crew set out on the voyage, floating with the current, but under the charge of a competent pilot, who is supposed to be familiar with the channel to Pittsburgh, as well as every rock, and bend, and eddy throughout the whole course of the river. During moonlight nights these craft "run all night;" but when it is dark and the weather stormy they generally "tie up" at some well known eddy, and wait until morning. These craft do not often find their way back, being converted to other uses, and sometimes broken up for the lumber they contain.

But the oil trade has given birth to some new and original inventions. These have grown out of the necessities of the case. Barrels are sometimes scarce, and latterly always high in price. The river is often too low for steamboats, and as something must take the place of these larger craft, some inventive genius suggested, and carried out the suggestion, of constructing what are called bulk boats. These are rude affairs, but answer a good purpose, They are made of two inch plank about sixteen feet square and from two to three feet in depth; divided internally by planks into bulk-heads of perhaps four feet square, to prevent any undue agitation of the contents by the motion of the boat. Sometimes these bulks are entirely decked over, so that the boatman walks upon the top to manage his boat. These boats sometimes have the oil pumped into them at the well, and are run to Pittsburgh without further expense than the cost of running them to market. When the oil is pumped from them they are broken up, and the material used for kindling wood, the process through which it has gone adapting it eminently for that purpose.

After this, there is a rude nondescript that surely was never dreamed of outside the oil region. It consists of a series of rough ladders constructed of tall saplings. These ladders are moored in the water, and between each pair of rounds is placed a barrel of oil, floating in the water, but kept in position by its hamper. A number of these ladders are lashed together, until the float contains several hundred barrels of oil. These unpromising fixtures, as well as the unwieldy bulks spoken of are often, during favorable weather, run to Pittsburgh with entire safety. To assist in this crazy kind of navigation the services of the old lumber pilots are called into requisition. Their training, in the days of lumbering and running pig metal, had rendered them thoroughly familiar with the river; but this trade had failed, and many of them were distressed that their occupation was gone. The advent of this new trade, then, was hailed with joy, as it afforded the opportunity of revisiting their old haunts again, and of facing the storm, and navigating their craft through the dangers once more.

A peculiar institution in connection with water navigation is the "pond freshets" in Oil creek. It has been quickened into life by the necessities of the business. The wells extend along the creek for a distance of fifteen miles from its mouth. The yield is enormous, and, as yet, land carriage is difficult and expensive. The valley is narrow, and the stream tortuous, and in many places the ground soft and yielding; added to this teams are obliged to cross and recross the stream continually. The creek itself is too small for navigation, except at occasional times during natural freshets. A railroad has been proposed down the valley from Titusville to Oil city, but the enterprise is of doubtful utility, as there are difficulties in the way that at present seem insuperable. The narrowness of the valley renders every available foot of land valuable for boring purposes, so that it would be difficult to locate a road that would not come in conflict with some individual's pet boring spot. The danger from fire would, however, be the chief difficulty in the way, as columns of gas are rising continually from the wells, which might at any time change this rich valley to a river of fire and death.

To compensate for all these difficulties a system of artificial navigation has been adopted. Before the advent of the oil business there was a considerable trade in lumber down Oil creek and its tributaries, and many saw mills were in operation that were driven by water-power. Most of these mills had fallen into disuse. Timber had become scarce, and there was no longer the same

Pond Fresh Smash Up, Oil Creek.—*Page* 165.

demand upon them. But the dams still remained, and were now pressed into a new service, of which their original builders never dreamed. Perhaps in all there are five or six of these dams, constructed with draws in the centre so that they can easily be opened at the proper time. By means of these dams, the water is collected and retained, so that "pond freshets" are arranged about two days in the week. The day and the hour are arranged beforehand for these artificial floods, and the oil men have every thing in readiness. At the appointed hour the upper dams are opened, and then as the flood pours down others below them give way adding to the volume, until the miniature tide has increased to a river. At each landing it receives its tribute of boats, until as the fleet approaches the mouth of the creek it numbers often times over two hundred boats, bearing with them from eight to ten thousand barrels of petroleum.

The advent of this fleet of oil boats at the mouth of the river is in the highest degree spirited and exciting. As boat after boat rushes into the river, there is the dashing to and fro of the boatmen, rapidly handling their sweeps, to avoid running ashore on the one hand, and against the piers of the bridge on the other. Sometimes the danger is from Scylla, and sometimes from Charybdis, and sometimes it is received from both in quick succession. Men are shouting their orders on board the boats, and multitudes, who have collected on shore as spectators, shout their applause in all directions until the excitement becomes intense.

Here and there a collision occurs, that often results in the crushing of the feebler boat and the indiscriminate mingling of boatmen, fragments of the broken craft, oil and the fixtures of the boat in one common ruin.

> "Apparent rari nantes in gurgite vasto;
> Arma virium, tabulæque et Troia gaza per undas."

A heavy oak barge running into a frail bulk constructed of pine planks, will pierce it as though its walls were simple paper, or two such barges will often crush the feeble little flat between them as they would an egg-shell, when the boatmen are forced to take an unwilling bath in the water, and sometimes in the petroleum itself. Often two or three boats are thus wrecked and the contents lost at a single pond freshet, involving, of course, a serious loss, yet it is one of the risks of the business that must be placed as an offset to greater gains in other directions.

In this fleet the form and variety of boats beggars all description. Sometimes there is the orthodox flat-boat with iron-bound barrels, with a show of respectability around it, and disposed to put on airs like a well-dressed swell in the midst of a crowd of ragged loafers. Next will follow a rude scow, and close upon it an unwieldy bulk, into which the oil has been pumped at the well, and, perhaps, bringing up the rear an unmanageable ladder-float, although these latter have lately been ostracised from the creek from their disposition to inflict damage and shipwreck upon the more respectable class of boats.

This extemporized navigation is kept up and regulated by a kind of code of honor. Written laws and legal enactments have not yet learned of its existence. By a mutual understanding each oil producer along the creek pays a share of the expense in proportion to the amount of oil shipped. This is at the rate of about five cents per barrel. Before a pond freshet is to come off an agent visits the wells and collects this amount from those who propose availing themselves of its benefits, and in this way the labor and care necessary to keep the matter in order is compensated.

After the oil produced by this pond freshet reaches Oil city, it is in part shipped down the river to Pittsburgh, in a manner already described, and in part is sent in other directions. River navigation is tedious, and for shippers to New York, Philadelphia, Boston, or the West the route is a very circuitous one. Better facilities are therefore looked for. Besides a large portion of the year the river is not navigable. Sometimes in the winter it is ice-bound for months, while in summer it frequently becomes so low as to be fordable, when the smallest craft must be laid up from use.

Soon after the business began to enlarge and prosper, a new link in the great railroad system of the country was completed that at once promised relief. This was the Atlantic and Great Western Railway, extending from Salamanca on the the Erie Railway on the East, and connecting with the Western roads in Ohio, forming a connection with New York on the east and the Mississippi and beyond on the west. This road passed within twenty-five miles of Franklin, and the idea was suggested of building a branch road from Meadville to the latter place. The project was a feasible one, inasmuch as the route lay immediately down the valley of French creek, the portion of which was already graded for the tow-path of the old Franklin canal.

But withal the company approached the enterprise with great fear and trembling. There was plenty of petroleum just then; but who would insure its continuance? The wells were yielding satisfactorily, but the bottom might, and probably would, fall out of them very suddenly, and then where would be the trade. The fear was before them that in the course of events the time might arrive when the road would be left naked and bare like the old dismantled oil derricks that were to be seen so plentifully along the banks of French creek, as a warning and a terror to all railroad companies, too much given to doubtful enterprises.

The enterprise was thought over, however, and inquiries put afloat as to whether Venango county was possessed of coal or any thing else that would afford a living trade when petroleum gave out. But these inquiries were bootless, for Venango then, as now, could only point to her oil wells, and cry out with the Roman matron, as she exhibited her children, "Behold, these are my jewels!"

Notwithstanding, the branch was completed, and soon assumed, not the place of a branch but the main trunk, both in freight and passengers; actually shipping as high as two thousand barrels per day. This road is on the broad guage principle, and, although the Franklin branch is a little slow from peculiar circumstances, yet the main trunk is admirably arranged and managed, and withal one of the most comfortable roads to travel upon in the United States. The Franklin branch was first built from Meadville to Franklin, but has recently been extended to Oil City. Before this extension was made oil was brought in boats from Oil creek to Franklin, and then shipped on board the cars. When the roads are favorable, an immense quantity of oil is still drawn in wagons from the Oil creek valley to Franklin. It is brought from the wells along the creek for six or seven miles above the mouth, making from twelve to fifteen miles the whole distance. As many as three hundred teams are engaged in this business, lining often the whole distance from Oil City to Franklin. The load for two horses depends on the condition of the road; but as it is down grade, it is usually from five to seven and even eight barrels. The price of hauling, too, varies with circumstances, as the distance, condition of the roads, and the activity of the market. Generally for the distance of twelve miles, the price paid is about two and a half dollars per barrel.

A railway car usually contains from fifty to sixty barrels, and the freight from Franklin to New York by this route is about three dollars. At Corry, forty miles east of Meadville, the Atlantic and Great Western railroad crosses the Philadelphia and Erie, where freight can be transferred for the Philadelphia market; but as the guages are different, there must be a trans-shipment from the cars of the one to those of the other.

Another mode of transportation, is from the upper portion of Oil creek by way of Titusville. The Oil Creek railroad extends from Corry, (where the Atlantic and Great Western and Philadelphia and Erie railroads intersect,) through Titusville to the Shaffer farm, five miles below. By this route a larger amount of oil has probably found its way to market than by any other. The shipments in 1864 amounted to 284,415 barrels. It is in contemplation to extend this road down the valley to Oil City, with what success this plan can be carried out, remains to be seen, as the obstacles in the way are very great. It would add very much to the convenience

15

and vigor of the oil trade if it could be done, and would be most advantageous to all parties concerned.

There are some other railroad projects on foot that will no doubt be carried out with success. The Franklin and Jamestown link is already in process of completion, and will, no doubt, soon be opened to trade and travel. It is to connect Franklin with the Pittsburgh and Erie Railroad at Jamestown, Mercer county. This will not only shorten the route to Pittsburgh, but open a new route to Cleveland—no mean competitor with some of her Eastern sisters in the oil trade—as well as other points in the West.

Another project still, is to continue the Allegheny Valley Railroad, already completed to Kittanning, from this point to Franklin. The distance to Pittsburgh by this route will be about one hundred and twenty miles; that by the way of Jamestown, about one hundred and thirty.

Other routes are suggested, but they are too far in the dim distance to be spoken of confidently or particularly. As the business enlarges and develops, new routes will be projected and opened as the exigencies of the time and occasion may demand.

In regard to the place where a market is sought, circumstances usually determine the destination. New York, of course, must always carry off the palm, from her unequalled facilities for exporting. It is the great centre of trade, too, and it is but reasonable that it should bring its advantages to bear in this important trade. Boston, being the first of the Eastern cities that entered heartily into the business, is still holding its own, as its enterprise and liberal views entitle it to do. Philadelphia, being of more sober and deliberate character, was dilatory in its efforts, but is attracting much of the business in its own direction. To Pittsburgh naturally belongs a large portion of the trade and influence, as well from its proximity to the oil region as to the convenience of transportation. It is still doing a fair share of the business legitimately belonging to the petroleum trade. Cleveland also comes in for a fair proportion of the shipments of oil.

The location of the companies that are operating influences to a certain extent the direction of the product. Sometimes companies operating in the oil region, are engaged in refining at home, and in this way are induced to bring the product of their wells to the location of their refineries. This is especially the case where the product is not extensive. A Cleveland, Buffalo, or Pittsburgh company will naturally ship toward their own centre of influence, and in this way the interests of many parties are subserved. The influence of capital and enterprise, as well as the demand for labor, is disseminated throughout the country.

Still a large proportion of the product of the oil wells is uncontrolled by influences of this kind. Where there is a large quantity of any valuable product it naturally and unfailingly seeks the best market. And that is the best market, other things being equal, where the best price is obtained, the most ready and constant demand, and the best facilities for reaching it. In these respects the seaboard cities must, of course, be pre-eminent, as the foreign market is now, and must still continue to make demand upon the production.

As to the first change of hands, great diversity prevails. Originally almost all oil was sold at the wells, and this, too, very soon after its leaving the regions below. The necessities of early companies and operators required them to dispose of their oil as soon as possible, but as the business has gradually assumed a new form, and larger capital is investing in it, this necessity no longer exists. Many companies ship their own oil to the sea-board, and even to foreign ports. There is still, however, much changing of hands in the movement of oil. There are parties who purchase at the wells and sell at Oil City or Franklin, and ship on their own account. There are oil brokers, likewise, at these points that buy to sell again, with no intention of shipping; so that the business is diversified in its character. The whole tendency, however, is towards centralization. Capital will have its influence, especially where that capital is accompanied by enterprise and energy, as is the case pre-eminently in this business. It is, perhaps, a foregone conclusion, that as time rolls by, the influence, as far as money is concerned, of this trade will be transferred to the cities, and less felt in the immediate region where the wealth is developed. The labor is small at best, and as facilities are multiplied for developing and carrying to market, less labor will be required, and less expenditure involved in transferring the product to the great markets in the East.

## CHAPTER XIII.

### FRANKLIN, FRENCH CREEK, AND SUGAR CREEK.

It was not long after the fortunate strike of Colonel Drake on Oil creek, before the matter of boring was suggested in the town of Franklin. When the news of the first success was spread abroad, the feeling expressed was that of willingness to be a partner or even a relative of the successful oil miner. But soon the idea was advanced that other efforts might be equally successful. Men began to cast about and inquire for oil springs and places where the oil exuded, even in small quantities, from the rock. In Franklin these indications were numerous and unmistakable. Some were in the wells of water, some in the bed of the creek, and some near its banks, giving tokens that were most encouraging and inspiring.

Companies were organized in an informal way, large and cumbrous, but the plan was to develop the resources of the place. In the meantime many a sweet golden dream was privately indulged in, that would prove a glowing reality after a while, when the matter should be developed. Had not the first oil well yielded forty barrels daily? Was not each barrel worth forty dollars? And why should not the same success attend operations in Franklin, with a much better "surface show" than where this first great success had been achieved? And would not the dividend, even in a company of ntty, oe a handsome sum from such a well?

Operations were commenced, sometimes by large companies, sometimes by small ones. Sometimes these companies were formed by citizens of Franklin, and often by strangers, until it seemed that the town was effectually aroused from the long sleep that had been almost like that of Rip Van Winkle in Sleepy Hollow. In the course of the summer succeeding the first successful experiment on Oil creek, there were not less than two hundred wells in different stages of progress in the town of Franklin alone. Wells were being bored in gardens, in door-yards, and even in some cases in the bottoms of wells from which water had been procured for household purposes. So numerous were the tall derricks, that a profane river-man made the remark that the people of Franklin must be remarkably pious, as almost every third man seemed to be building a "meeting-house" with a tall steeple near his dwelling. At one time there were in Franklin fifteen productive wells, yielding a daily aggregate of one hundred and twenty barrels.

The first well that was found productive in Franklin has become historical. It has been placed in the annals of petroleum history, and will go down to succeeding generations side by side with other "first things" that have become immortal. This well was known as "the celebrated Evans well," and was the third successful well that was bored—Drake's being first, and McClintock's second, both on Oil creek.

It was, in some respects, the most remarkable well in all the region. It was sunk by its proprietor in the bottom of the well that had long been used for household purposes. He was induced to choose this spot from the fact that a trace of oil was generally found on its surface, particularly in summer, when the water was low. Another thing that induced him to think that oil would be found on his premises, was the fact that on the bank of French creek, that was skirted by his lot, a small oil vein was constantly yielding a limited supply, which had been sufficient for the wants of his family and of the neighbors. Times have changed since then. The wants of the people are greater now than formerly. It requires a flowing well to satisfy them.

An humble house and lot constituted the entire worldly wealth of this worthy man. The work in the well, too, was performed entirely by his own family. Being a blacksmith, he constructed his own boring implements, as well as kept them in order, and was dependent on no outside assistance. Patiently and assiduously did the blacksmith and his sons toil on, as they had seldom toiled before, the former guiding the drill, and the latter applying the power by hand, in the use of a very simple spring-pole arrangement. All wished for their success, but many doubted it. The old salt-borers who were operating elsewhere looked wise and shook their heads. His simple apparatus and want of experience would certainly bring disaster and disappointment;—so they argued. It was "red tape" against honest yet unsystematic effort.

Thus they went on until, at the depth of only seventy-two feet in the rock, they struck a crevice that promised to pour them out rivers of oil. In attempting to enlarge this the drill broke, the fragment remaining in the cavity, and, for the time, resisting every effort used for its removal. The well was then tubed, and a hand-pump inserted, when it was found to yield at the rate of ten or fifteen barrels per day. The labor of pumping was great, however, and the work was not kept up continuously. The excitement attending this development was very great. Crowds visited the well daily, and an almost constant levee was held around the pump and tank that received the oily treasure. Speculation soon began to run wild, and the fortunate owner of this well, among other propositions, received an offer of fifty thousand dollars for his well. Here was treasure, such as he had never dreamed of, placed within his grasp. But it was of no avail. To all these tempting offers he persistently made the same reply—that he had bored that well for his own use, and that if others wished a well, they could do as he had done; the sites were numerous, they could bore wells for themselves.

Oil was generally found around Franklin at the depth of about three hundred feet from the surface for pumping wells; in the valley of Oil creek the same stratum was reached at about half that depth, intimating that the rock had a dip or inclination to the southwest. In all these attempts, whether successful as oil wells or not, a strong body of salt water was obtained, that added greatly to the facility of separating the oil, by its increased gravity. Hitherto the business had been pursued with encouragement and profit, to those who were engaged. The demand was steady, prices remunerative, and visions of untold wealth were looming up before the minds of operators, when the shock came that shut up at once all small wells, and suspended all further operations. This was the opening of the flowing wells on Oil creek that brought down the price of oil, so that it was not worth the coal necessary to pump it from the rock.

Here was another panic. Previously there had been a general stampede towards the oil regions, and Franklin had been the Mecca from many directions; now the stampede was the other way. Wells were abandoned that had been prosecuted with encouragement, and leases forfeited without an effort to retain them against better times. Lots that had been purchased at high prices were sacrificed. Instances can be pointed out where the disappointment and disgust were so great at this new feature, as to lead not only to the abandonment of the lease and derrick, but to the boring implements themselves. In one or two instances they remain in the dismantled derrick until this day.

But worse than this. In many instances companies owning wells that were yielding from three to six barrels, removed their machinery, abandoned their leases, and left the country. They thought the days of small wells were numbered, and that they could never be successfully worked again. During the last year some of these abandoned wells have been yielding from seventy-five to one hundred and fifty dollars per day. By this untoward course of events, vast sums of money, gathered all over the country, were sacrificed, much of which might have been saved with care and foresight. But the history of the oil business thus far has been but a succession of different kinds of excitements. Men are not content to make money slowly and by patient waiting in the oil business, as in other walks of life. Wealth must come at once, or they are discouraged and disheartened, and abandon the field.

The region around Franklin and the flats of French creek has never been fully tested. It was just in the full tide of boring, and before any one had reached the third sand rock, that the discouragement, growing out of the opening of flowing wells came upon the business here, and all effort ceased at once. In Franklin these efforts have not been resumed; with the exception that several wells are now successfully pumped. In addition to this, there was another ground of discouragement. The theory got afloat on the public mind, that no oil would be found in the sand rock or below it; and very few had the courage to go below three hundred feet. If no oil was discovered at that depth the general disposition was to abandon the search, and try some new location. At this depth many wells yielded a few barrels a day for a time, and then run down to one, or ceased altogether. This induced many a hopeful operator to settle down on the opinion that Franklin was rather on the outskirts of the oil region, than in its central basin.

There is another feature of the operations here that induces to the belief that this region has not been fairly tested, and that is the character of the oil produced. It is uniformly a heavy oil, from 30° to 33°; whilst that from the flowing wells on Oil creek is from 40° to 50°; implying that Franklin and French creek oil is of a surface character, and has found its way up through seams and fissures from the deeper oil strata beneath. The fact has already been noticed that the general inclination or dip of the rock strata is toward the southwest. This would lead to the supposition that when the third or fourth sand rock is pierced here, flowing wells may be discovered that will amply reward the persevering experimenter.

The character of the oil in Franklin and on French creek adapts it rather for lubricating than refining purposes. It was, at first, rather undervalued, as not yielding a large per centage of illuminating fluid; but soon began to rise in the market as a valuable lubricator, being used for many kinds of machinery without any preparation whatever. It now commands a ready market at prices more than double that of the deep wells of Oil creek. To the operator this is a great advantage. The saving in handling, in barrels, in freight, and other things add very materially to the margin of profit on even small wells producing heavy oil.

French creek is one of the important tributaries of the Allegheny, emptying into it at Franklin. Its general features would indicate a geological structure of the country favorable to the development of petroleum. The valley on the eastern side has considerable width, and the banks, in places, present a rough and fractured condition of the rocks. It has moreover, from time immemorial, been noted for its "surface indications."

Like Franklin it has had its discouraging experiences. The tide of imperfect experiment swept down the valley in 1860, leaving behind it little except blackened derricks, the remains of shanties, and the shattered masonry that told where engines had been set. Still a number of successful wells were opened during the first year of active operations, which have been producing regularly and faithfully to the present time. There are now (March, 1865,) on the banks of French creek some six or seven producing wells yielding oil of some 31° in gravity.

During the past summer almost all the lands on French creek, for a distance of twelve miles above Franklin, have been purchased and placed in joint stock companies, and every means that capital and enterprise can bring to bear will be expended in their development during the next season. A deep well, sunk in the neghborhood of the old forge, produced a very thin, light colored oil, indicating that, under favorable circumstances in developing, this region may yield largely of the lighter oils when the proper depth is attained.

Patchel's run is a small stream that puts into French creek, about one and a half miles above Franklin, and runs in the general direction of Oil and Sugar creeks. At its mouth are some of the most productive wells on French creek. From the broken and precipitous character of its bed there is ground to believe that at some future day it will prove to be one of the desirable regions for mining. Its lands have already been brought into the market.

Eight miles from Franklin, on the western side of French creek, is an important tributary called Mill creek. It is a rapid, brawling stream rushing down amid the hills, with here and there indications that amidst its broken rocks the treasure is but biding its time, and will ere long manifest its presence. The lands along this stream have been nearly all purchased for oil purposes, and will ere long be developed. At the mouth of this stream Utica is situated, a thriving village, hopeful and full of enterprise. Two and a half miles further up, is another stream called Deer creek, coming in from Mercer county, and is now attracting the attention of oil men.

Sugar creek is an important branch of French creek, joining it about two miles from Franklin. It runs in a northerly direction through some of the finest portions of Venango county. The Sugar Creek valley has always been celebrated for its fine farms, and thrifty farmers. In this respect it affords a striking contrast to the territory along the Allegheny and some portions of Oil creek. In the latter places the doctrine of compensation seems wonderfully illustrated; in the former, if it should ultimately prove to be good oil territory, the evidence of the goodness of Providence would seem to be cumulative.

It is but recently that much attention has been paid to this region as an oil producing section; still as long as three years ago a well was put down by Joseph McCalmont, on his own premises, about two miles and a half above the mouth of the creek. This well was bored by water-power, taken from a small spring run, and with very imperfect machinery. It was bored only about three hundred and twelve feet deep, reaching the second sand rock. During the three years some three or four hundred barrels of oil have been drawn from this well. It is of a dark, heavy character, and at the present time much sought for, in the market, as a lubricating oil. The well has only been worked at intervals. Sometimes the water was not sufficient, as in times of drouth, and sometimes, as after the flowing well panic, the price of oil would not justify the proprietor in pumping even by water-power. The probabilities are, that if an engine was employed, and the operation of pumping kept up steadily and perseveringly, a larger yield of oil would be the result. As it is, a yield of four to six barrels daily would be a more productive source of income than the best farm on Sugar creek.

From this well up to Cooperstown, some two miles, preparations are making to test the rocks effectually. Many wells are going down with encouraging tokens of success. In the neighborhood of Cooperstown a well has been going down with constant evidence of the presence of oil until it has attained the depth of six hundred feet. What the ultimate result may be cannot be told; but the constant, universal presence of oil in this valley, whenever the sand rock has been pierced, and even the stratum above, must be evidence of its existence in large quantities below.

Recently a new well has been opened on Homan's flats, about three miles from Franklin, that has brought Sugar creek into the very vortex of the excitement. This well is about three hundred and fifty feet in depth, and was not at all promising at the first tubing. There was oil, but not the rushing flow of the Oil Creek valley. It was tubed and the pump put in operation, but for some two weeks operations were very discouraging. The yield was a steady stream of water; but, at the end of two weeks froth and foam began to appear, and then oil, until, by steady improvement, it yielded from fifty to sixty barrels per day. The vein pierced was, no doubt, connected with a distant oil cavity, and it required long and persistent pumping in order to exhaust the water.

This new feature is most encouraging to operators who are commencing at a distance from large wells, and has been the means of calling attention to territory previously under valued.

# CHAPTER XIV.

## OIL CREEK.

As yet Oil creek is the most important region in all the Oil valley, or in all the wide world. Nor is it an unreasonable supposition that it will so continue. If we take into consideration the antiquity of the developments along this creek, the magnitude and splendor of its resources as brought to light during the past few years, and the unabated richness of the product in the present, we are safe in the conclusion that here nature has lavishly provided for the wants of the present and coming generations.

This region was first in the field as oil territory. Here the great oil springs were worked with profit, and the first well bored in the rock. To this region the first attention was directed from abroad, as the value and importance of the product began to dawn upon the country. It was but natural that attention should be attracted first to the new well, as though its neighborhood was the only place where oil might be expected. It was natural, also, that the next thought should be, that in the vicinity of the lower oil springs on the McClintock farm, success equally good might be obtained, and then in general that wherever oil springs were found at the surface, oil might be expected in the rock below. It was not strange, then, that the principal attention was directed to Oil creek both at home and abroad.

The success of the first efforts on Oil creek, too, encouraged all comers to seek locations beside new wells that were yielding oil. The first two wells that were bored were successful. Drake's well, yielding from ten to forty barrels per day, and McClintock's, lower down on the creek, yielding still more, induced the expectation that wherever the rock was pierced to a depth of seventy feet there would be a successful well. This early prestige gave this region a popularity and an importance that it has never lost. It was recognized as the child of fortune, petted and spoiled somewhat, but still in boundless favor.

But another reason for the early favor that was shown to this region, was from the name the early fathers had given this stream. It was a proper name, a significant name, and one to which it was richly entitled, and yet it had a persuasive power that was boundless in influencing early operators in selecting a site for their operations. True, authority, to which we all yield in the general, is somewhat against this idea:—

> "What's in a name? that which we call a rose
> By any other name would smell as sweet."

Yet it is somewhat doubtful whether, if any other name had been applied to Oil creek, it would have appeared so unctuous and so promising for the pursuit of the petroleum business. The natural reflection of most persons would have been, that they were seeking for oil, and what place or valley so favorable for this pursuit as the valley of Oil creek. And so the multitudes sought this valley, assured that it was the best possible location for boring wells.

And there was reason for selecting this above all other places for the business in hand. Here oil had always manifested itself; here were many successful wells; here was OIL creek, above all other names the one to encourage and cheer, during the long days and weeks when the drill was forcing its way downwards.

And this region has never disappointed the expectations that were formed concerning it. It has uniformly proved true to its ancient fame and time-honored name. If there have been failures, there have been unbounded successes. If many a man has left this region poorer than when he came, his experience has been the exception and not the rule. Colossal fortunes have been made here in a short time, but yet fortune is not designed for all men, and many fail of realizing their expectations even here.

The natural features of the valley of Oil creek have great variety. Near the mouth, the valley has expanded so as to form quite a little extent of flat land, and, in former times, a single row of farms extended up the valley that were quite productive. From here up, the land is more broken, and in places the hills on either side approach each other, leaving but a small extent of flat land, and again the hills recede, forming a wide valley, with the creek sometimes on one side and sometimes on the other, until Titusville is reached, fifteen or sixteen miles above the mouth of the creek.

The stream is very tortuous, winding in every direction, and rendering it necessary that a passage up or down the valley should be one continued series of crossing and recrossing its muddy, oily current. The hills on either side, though generally steep or rather precipitous, are usually covered with soil that sustains a growth of bushes and trees. It is very seldom that the rock crops out from the face of the bluffs. These hills are frequently cut on either side by ravines, through which streams of water find their way to the creek. Some of these streams, although disregarded at first, have proved quite as valuable for oil purposes as the most notable portions on the creek, and this, too, far up above the level of the valleys.

The general productiveness of the Oil Creek valley varies considerably. It is not scattered over its whole extent that we find those enormous wells that seemed at one time as though they would deluge the whole land, but generally congregated together in particular localities. There is one region, of about four or six miles in extent, that resembles the milky-way where the stars seem massed and wedged together; yet, withal, unlike the milky-way in another respect, for they are stars of the first magnitude. This is in the region of the McElhaney, Foster, and Hyde & Egbert farms, where we find congregated the Funk, Sherman, Empire, Maple Shade, Noble, Jersey, Coquette, and other wells, a grand and noble constellation, that have yielded their products most lavishly.

Still this is no indication that other portions of the land along the creek may not prove equally valuable. Portions of land that were once denominated the "Dry-diggings," as the Story farm, have proved as valuable as any other portion, whilst other portions, once the favorite resort, have, for a time, lost their popularity, to regain it by new and more successful developments.

The flowing wells, that have made the business so famous, were first opened in the valley, and are still a most important feature connected with it. Although the eariler ones have ceased to flow, yet new ones are taking their places from time to time, so that the supply is kept up with considerable regularity. In the meantime, many of the wells that flowed until the yield was absolutely enormous, assumed the form of pumping wells, and are now worked successfully in this way. Other wells are opened that have not the natural characteristics that cause them to flow, and yet yield plentifully by the use of the pumps. Sometimes these wells interfere with one another, though this is not a frequent result. It is a bare possibility for two parties to strike the same vein, and in this case the one nearest the grand cavity will have the advantage. But this does not often happen, even where the derricks are planted as thickly almost as the trees in the forest, because these veins, as a general thing, seem to run vertically rather than horizontally. No doubt they run in every direction, yet their general course is vertical, leading from the great reservoir below towards the surface, ramifying as they proceed, and, of course, becoming smaller until they reach the upper stratum of rock, and some of them even the surface. If we take this view of the nature of the structure below, we may well suppose that the cases of interference will be few in comparison with the whole number of wells. And the history of oil operations, where numerous wells have been put down in near proximity to each other, bears out this view of the case.

The whole number of wells producing oil is not easily estimated, as many are productive, or may be rendered so by proper care, but are remaining quiet from various causes. Sometimes there is a conflict in title, or machinery is defective, or parties are called away from various causes. But the number is steadily increasing, and the supply of oil kept up. This supply has been rather oonstant for the last three years. If the business has grown dull through low prices and the scarcity of fuel, it has been counterbalanced by the advent of new flowing wells; and when some or many of them have ceased flowing, a number of new pumping wells have supplied their places, so that the supply has nearly kept pace with the demand.

As to the depth to the rock and the dip of the strata in Oil creek there is quite a variety. The greatest depth of soil overlying the rock is on the flat or low ground, while the rock is but slightly covered, and even exposed, on the hillside, at the foot of the bluffs, and at the base of the hill opposite to a flat. This arrangement is almost invariable. On the "Blood farm," which is, perhaps, six miles above the mouth of the creek, the rocks are exposed to view at the base of the hill, which is also the water's edge, while on the opposite side of the creek, where there is a broad flat, the operators are obliged to dive pipe thirty-five feet before reaching the rock, showing evidently the indications of an upheaval of the rock at the base of and underneath the hills.

And here, as elsewhere, the base of the hill generally produces the most successful wells, although this is not an invariable rule. Higher up the creek on the Story farm, the broad flat has, as yet, produced very little oil, while at the foot of the bluff there are many fine wells. On this farm the flat is called the "dry territory," and the foot of the hill the productive territory. There is a geological theory made good in this experience. The upheaval, of which we read on the very first page of the rocky volume, as it is turned up under the base of the hill, has broken, and shattered the rocks at the base, leaving the strata inclined at all angles, and, of course, full of cavities and even huge caverns, into which the oil, as a matter of course, collects from the smaller apertures, leaving the unbroken strata in the neighborhood with little or no resources upon which the operator may draw. The result is that wells that are sunk at the base of the cliff, are located immediately over this shattered, cavernous condition of the rock, and reaching some of the crevices or leaders that connect with the crevices below, yield luxuriantly; whilst those sunk in the flat that has not been disturbed so much reach only insignificant veins, that abound only in water, and the pump not having sufficient influence to draw the oil from its distant bed, the well is a failure.

On the Blood farm the nature and location of the strata will convey a good idea of the general geological features of the rock in the valley. Out on the flat there is thirty-five feet of superincumbent earth before reaching the rock. The first rock is quite hard and even siliceous for the distance of fifteen feet, where it changes to the ordinary shale. At the depth of two hundred feet the first sand rock is reached. This is from twelve to fifteen feet in thickness. The second shale is about one hundred feet in thickness. The second sand rock from fifteen to twenty feet thick. Third shale one hundred and fifty feet, and third sand rock about twenty feet. This is a register for a well bored through the third sand rock, and is something over five hundred feet in depth. It will be perceived that the upper stratum of shale is much the thickest. The first sand rock is not so uniform in thickness as is the second and third, nor is it quite so thick.

On this part of the creek a well is not considered finished until it has passed through the third sand rock, as the largest and best wells have been found in this stratum. Nor have many persons, thus far, gone below this, save to prepare a receptacle for the deposit of sand and sediment in the working of the well. Still a few have bored to the depth of from seven to nine hundred feet, with various results.

The deepest well yet bored in the oil regions is on Watson's flats, two miles below Titusville. This well has been bored to a depth of twelve hundred feet, and has not yet reached the third sand rock; showing that there is a dip in the rock towards the north, indicating that there is on the creek an anti-clinal axis, that accounts for successes on the central portions of its territory, and failures above.

As to the comparative success attending the boring of wells in the Oil Creek valley, the history so far would indicate that about one well in ten is a success. What is termed a successful well is one that can be worked with profit to its owners. No doubt, as stated in another place, many a well that is abandoned as worthless, might be made to yield by proper treatment in tubing and pumping, and many others might be rendered profitable by deeper boring; but to take results as we find them, the proportion stated above will perhaps be found nearly correct. This leaves a large margin for the cultivation of faith and the exercise of hope, whilst engaged in the operations of boring and tubing. Yet there is this encouraging feature connected with the business generally, that there is no greater proportion of failures now, that the valley is pierced like a colander, than there were at the beginning of oil operations. No doubt the great number of wells that are drilled through the rock will have a tendency to lessen the number of flowing wells by gradually withdrawing the gas, but it yet remains to be seen whether it will relatively lessen the amount of oil yielded by the process of pumping.

During the first year of the oil business, the choice locations were supposed to be near the water's edge, but generally, as lots were leased and lands occupied, new comers were forced back to the base of the hill. Experimental boring has proved equally as satisfactory there as at the edge of the creek, and even more so. As land became scarce, derricks were seen planted on the steep edges of the bluff, quite a distance above the level of the creek, and still the rock was found prolific, the only inconveniences being the difficulty of bringing fuel to the engine, and finding a location for oil tanks and barrels.

In the meantime little tributaries that flowed into Oil creek on either side, which had been neglected before, began to assume importance. Some of them were little dancing brooks that almost leaped down the cliff side, but the theory began to be entertained that probably the strata of rock would rise with the acclivity of the bed of the stream, and oil be reached there as in the valley below. Practical results have shown that this theory is correct.

Cherry run is a most important branch of Oil creek as regards the operations connected with it, although it is small and insignificant as a stream. It enters the creek on its eastern side near the town of Rouseville, and bids fair to become as famous in history as many another trifling little stream that has been associated with heroic deeds or important achievements. It is a small, impetuous brook that, near its mouth, leaves scarcely room at its side sufficient for a wagon road. Following its banks for the distance of five miles we find the little town of Plumerville, on the road leading from Franklin to Warren. Here very extensive operations are carried on connected with refining and preparing oil for market.

The valley of Cherry run widens and enlarges from Plumerville up, and is now a scene of bustle and excitement, as every portion of it from its mouth to a distance of ten or twelve miles is thickly bristled with derricks, and busy with the movements of the workmen. Nor is this mere speculation, or the result of excitement directed in this particular way. The successes that have attended experiments in this little valley are much greater so far, than those connected with the parent valley. For a time not a single failure occurred in boring on Cherry run, whether near to the current or up on the steep sides of the bluff that come near meeting on either side. Whether this uniform success was owing to the fact that the wells happened to strike the oil crevices in every case, or whether these crevices, with corresponding caverns, are more thickly located in this region cannot be told. Success in a given locality is the rule that governs its popularity, and the success of Cherry run territory thus far has established its reputation among oil men.

It is in this valley, some two miles from its mouth, that the famous Reed well is located, flowing some two hundred and fifty barrels per day. The proprietor of this well has had much experience in boring wells, and had learned "to labor and to wait." He had bored on French creek and on Allegheny river, but without success. His efforts seemed all to fail, and his hopes to meet with disappointment. Finally he resolved to try

Cherry run, and found fortune propitious. A prime vein was pierced, a flowing well was the result, and he can now repose on his laurels. Other flowing wells are in the same neighborhood, and pumping wells that yield bountifully are strewn thickly around.

A grand and magnificent view may be obtained by ascending to the top of the bluff in the neighborhood of the mouth of Cherry run. Not less than one hundred and fifty engines may be seen at a glance, all in active operation, with their columns of steam and smoke, running in every direction. The profusion of derricks is absolutely bewildering, resembling, more than anything else, a new town in process of building, with the simple frames of houses awaiting the roofing and weather-boarding. In connection with this the creek is seen filled with boats, and if a view of the river is obtained, it is seen swarming with flatboats and steamers. It is a busy scene, full of active life, and full of wondrous promise.

Rouseville is an active little village. Much business is transacted here in the way of selling and leasing lands. It was near this place that the celebrated "Burning well" was located, in whose fearful conflagration H. R. Rouse lost his life. The place takes its name from Mr. Rouse.

The next stream that enters Oil creek is Cherry Tree run, coming from the northeast, and joining the creek at the lower end of the Rynd farm. This run is divided into several branches; the first is Weikel run, and the next in importance Noell's run. Although no important developments have yet been made in these valleys there is nothing improbable in the idea that they will prove as productive and valuable as the other tributaries of Oil creek.

There are some other runs of minor importance, such as Corn Planter's run, Bennehoof's run, Fox run: these will ere long be heard from in the way of producing oil, that will bring them into notoriety, as the demand for territory is unceasing, and operations extending wherever there is the least prospect of securing oil.

Above Titusville, which is in Crawford county, numerous derricks indicate the intention of exploring thoroughly the whole region of the creek. There seems to be no reason why such explorations should not succeed, unless the geological structure of the rocks should be different from that of the region below. This would seem to be the case from the fact mentioned as to the trend of the rocks two miles below Titusville. Still as the face of the rock is frequently undulating, valuable territory may be opened up at no great distance above.

Titusville, like many other places in the oil valleys, has become decidedly famous for the last few years. It is situated in Crawford county, but as its influence and importance grew out of the oil wells, it is properly mentioned in connection with the history of Venango county. It is an old town; but, like Franklin, has taken a new lease of life, and is becoming rejuvenated once more. It is on Oil creek, and at a point very favorable to trade and business, being connected with the Atlantic and Great Western and the Erie and Philadelphia railroads by the Oil Creek railroad. The situation is good, and the streets laid out regularly. It has much beauty connected with its location and surroundings, and bids fair to become a place of great importance. From this point the railroad extends down

Oil creek to the Shaffer farm, and it is in contemplation to extend it to Oil City. In this way a much larger amount of trade will be secured, inasmuch as the present way of conveying oil to Titusville is both difficult and expensive.

Titusville, from a small village, has now assumed the proportions of a large town, increasing within a few years from a few hundred to a population of six or seven thousand. It has four churches, some of them of very fine architectural beauty and proportion, with an amount of taste that does great credit to the place. It has also two banks of issue and deposit, with a large number of hotels.

The banking interest in the oil region is one of immense importance. Besides those mentioned above, Oil City has two banks of issue and a third of deposit, whilst Franklin has the same number, making in all six banks of issue and two or three of deposit or exchange. All these have grown naturally and legitimately out of the oil business of Venango county, and are safe and highly prosperous institutions. This business has been largely indebted to the enterprise and foresight of Hon. C. V. Culver, who has been connected with it from the beginning of the prosperity of the oil region.

Oil City is an active, energetic place, full of hope and promise for the future. The site is not pleasant nor favorable to expansion, yet to lovers of the beautiful it is very picturesque. The principal part of the city lies along the base of a steep, precipitous hill, that would, in many places, be dignified with the name of mountain; on the other hand is the Allegheny river. There is at the lower end of the place room but for a single street; at the upper end, and across Oil creek, the valley widens out somewhat, particularly up Oil creek. The lower portions are mainly occupied for business purposes, and the level land does not suffice for this, but business is pushing its way into the hillside. On the northern side of the creek there is a beautiful section of the city, called Cottage Hill. This section is located on the steep hill-side, out of the way of fire and the terrible atmosphere that arises from the oil boats and wells, but particularly from the refineries that are in the neighborhood. From this point there is a fine view of the river, the town and creek, that extends to a considerable distance in every direction.

Oil City has already several churches organized, with church edifices, either erected or in progress of erection, and everything betokens prosperity, enlargement and wealth. There are two banks of issue and one of deposit, whose business would utterly confound the officers of many of the old city banks. The lower portion of the city is literally covered with oil derricks, encroaching even on building lots and dwelling houses. Some of them are in operation and yielding oil, others are just commencing operations. These wells, however profitable to the owners, will render the town rather liable to accidents from fire.

Across the river, and just opposite the mouth of Oil creek, is situated the new city of Venango. This, too, bids fair to become a young giant in the race for prosperity and growth. One year ago it was a farm, with a single dwelling house. Now buildings are multiplying rapidly. There lacks but one thing to insure the prosperity of Venango, and that is a bridge across the river. With this, the greater portion of the new population would select that place as their residence, whilst trans-

View of Oil City, Pa.—*Page* 196.

acting business at Oil City. The founder of this new town was William L. Lay.

The population of Oil City was at first floating and transient; latterly it is becoming more fixed and permanent. It has, probably, at the present time, from six to eight thousand. The other towns on Oil creek, as Tarr Farm, Petroleum Centre, and McClintockville, are all flourishing. In fact, it is almost impossible to tell where one ends and the other commences, so thickly strewn is this entire valley with a dense, active population.

The lands along the creek, between the same points, are estimated at a bona fide cash value of two hundred and fifty millions of dollars. Many portions of these lands have been sold at prices that would bring the whole at the same rate to this sum in the aggregate. The distance is but little over fifteen miles, and the valley narrow throughout its entire extent, so that an idea can be readily formed of the immense wealth contained in its bosom.

As yet the most extensive oil companies are found in operation in the Oil Creek valley, and much of the prosperity of the business is due to the efficiency and enterprise with which they have been managed. Among these are enumerated the "United Petroleum Farms Association," on the Graff-Hasson farm, near Oil City; the "Blood Farm Petroleum Company," on the Blood farm; the "Home Petroleum Company," also on the Blood farm, on the east side of Oil creek; the "Tarr Farm Petroleum Company," and the "Central Petroleum Company, of New York," on the Washington McClintock farm, with many others that are carried on with great efficiency and success.

As the names of Eveleth, and Bissell, and Drake are mentioned in connection with the history of the earlier developments of the oil business, so there are other names that belong to the history of the later development and successful prosecution of the trade, that must ever be associated with enterprise, and judgment, and perseverance. Among these are Brewer & Watson, formerly of Titusville, the brothers Egbert, Halderman, Shrieve, Brough & Tilson, Mitchell & Brown, Prentice, Clark & Seely, Coleman & Ewing, with many others, whose influence has contributed largely towards making the business what it is—the wonder, and pride, and glory of the whole land. The history of Oil creek would not be complete without these names.

There is a beautiful Indian tradition in connection with the early history of Oil creek that has a bearing upon the subject in hand. Cornplanter had lands here, and the valley was a place of frequent resort by the Indians, on account of its healing oil.

The tradition is, that many moons ago—long before the recollection of the most aged chieftains of their tribes—one of their bravest chiefs was afflicted with a painful disease, that was rapidly preparing him for the happy hunting grounds of the spirit land. For the good of the tribe he longed to live, and fasted and prayed to the Great Spirit to spare him until his tribe should be delivered from their difficulties. The neighboring tribes were on the war path, and he feared that his people would fall before them, and be scattered like the sere leaves of the forest.

The Great Spirit was propitious, and answered him kindly—

> "Spake to him with voice majestic,
> As the sound of far-off waters
> Falling into deep abysses,"

telling him that in the valley that should be pointed out to him he would find a great medicine, bubbling up through the ground and mingling with the waters, that should heal him of his maladies, and give him strength to smite his enemies and overcome them. The voice of the Great Spirit, moreover, assured him that this medicine fountain would continue to yield its supply until his tribe should cease following the wilderness and the war path, and be all gathered into the happy hunting-grounds of their fathers; and that it should then be given to a tribe of strangers, with pale faces, who should come over the big waters, and be by them desecrated to common and base uses.

The chieftain rose from the ground, and, although faint with fasting and weakened by disease, set out in quest of the medicine spring. The sun was setting, and the curtains of darkness were gathering around, but there was a light that glowed in the red chieftain's heart. From his lake-side home he turned his back upon the North star, and, faint and weary, he at last reached the place pointed out by the Great Spirit, just as the sun was rising in the east. The medicine was bubbling up with the water, the chief recognizes the gift, and finds healing and life in its power. The fountain has continued to yield its supply. It is still the gift of the Great Spirit, and its supplies should be received with gratitude.

# CHAPTER XV.

## ALLEGHENY RIVER TERRITORY.

THE region along the Allegheny has not been in course of development as long as Oil creek, nor has it been developed with such persistent effort and energy, yet it has so far more than realized the expectations that were indulged concerning it. The Allegheny is a beautifnl stream, well meriting the name the French bestowed upon it—La belle riviere. It is remarkable for the boldness and richness of the scenery found along its banks, and for the endless variety that characterizes its conformation. The current is winding, running in almost every direction, now sweeping by the base of the bluff on its eastern bank, leaving a little valley on the western, and now changing to the western side and washing the feet of the cliff in its clear cold waters. The hills are high and precipitous on either side, and, withal, the rock is very seldom seen cropping out from their sides. Here and there a gorge opens out between the hills, through which a creek or run finds its way into the river. Sometimes these gulches are dry, and show very little indication of having recently contained water. Oftentimes they appear to have been formed by some internal eruption, or convulsion of nature, as the sides are rugged and broken; they are to deep, too, to be the result of erosion. What is called the "Dry Hollow," some few miles above Franklin, is of this character. In driving a pipe, preparatory to boring a well, in this hollow, it was found to be over an hundred feet to the rock, as though there had been a mighty chasm opened between the hills. Whether this will prove a favorable site for oil wells remains to be seen.

The upper Allegheny, or that portion above Franklin, will first be noticed. A considerable number of wells are found on either bank, between Franklin and Oil City, all of which are worked by means of pumps. There are at present some fifteen wells now in active operation, averaging, probably, five barrels per day.

The "Two-mile run," so called from its distance above Franklin, was at one time considered a favorable locality for oil, and some three years since produced a considerable amount, but latterly little has been done in that vicinity. It has not yet recovered from the ruin and discouragement visited upon small wells by the flowing well excitement. It will yet, no doubt, be valuable.

Passing by Oil creek, in this place, we find Horse creek, a considerable stream flowing in upon the eastern side of the river, about ten miles from Franklin and three above Oil City. In this region the indications have been favorable, and signs propitious for many years. A Philadelphia company is operating here with favorable indications of success. The valleys formed by the river and creek at this point afford a large amount of bottom land that is favorable to the operation of boring.

Six miles further up, we come to "Pit Hole creek," now become famous for its new developments, as well as new features in the oil business. This is a rocky precipitous stream, dashing down amid the hills over loose, broken rocks, and has a fall per mile greater than any other stream in the oil valleys. It derives its name from a peculiar conformation of the country. The stream seems to gush from an immense pit in the rocks. There are several of these pits in the neighborhood that seem to indicate a structure of rock different from that known in the neighboring valleys. An old story is related of a hunter in the olden time, who had shot a wild turkey, and, wishing to go further and not carry his game with him, suspended it in one of these pits by a string. On his return later in the day, he found his game tainted and worthless. Others speak of villainous odors arising from these pits. The conclusion is, that they lead to broken cavities deep in the rock, and that gasses are escaping from below.

Be this as it may, Pit Hole has become a popular resort for oil men, and considered of the greatest promise for future developments. The best well on this stream is about six miles from its mouth, and has perhaps attracted more attention than any other well that has been opened during the last three years. While the well itself is not very deep, being but six hundred and six feet, it is located upon quite high ground and has passed through the strata that is usually found on the river-flat. It has even pierced the fourth sand rock that has never been reached on Oil creek. The region has not been leveled to ascertain the exact height; but this well cannot be less than four hundred feet above the level of the Allegheny, six miles distant. Here then is the sand rock upheaved to a level with the surface within comparatively a short distance. Might it not be supposed that underneath this region there would be a condition of the underlying strata peculiarly favor-

Flowing Well at Pithole.—*Page 203.*

able to the secretion of oil? If this upheaval was by volcanic action, it might very reasonably be concluded that the rock beneath would be contorted, and broken, and left full of extensive cavities, and, being in the oil region, these cavities would almost necessarily be filled with oil.

This Pit Hole well flows spontaneously at the rate of one hundred barrels per day, and this under great disadvantages, as the sucker rods are yet in the chamber.

A wondrous development is taking place in the resources of this region. It appears to be outstripping Oil creek itself, both in the success of the oil operations, and the growth of its population. The first well opened was the "United States," in the spring of the present year,—1865. This at once attracted the attention of oil men, when the tide of speculation commenced, and has continued with unflagging energy to the present time. There are now ten flowing wells, yielding from two hundred to eight hundred barrels each per day, on Pit Hole creek; with this encouraging feature, that so far there has been no evidence of decrease in the yield. Towns have sprung up as though by magic. Pit Hole creek has, in less than two months, attained a population of from five to seven thousand. The houses built in the same period of time number about five hundred. The place has already two banks, numerous hotels, theatre, and arrangements for religious service. Two railroads are in process of construction to the neighborhood of Pit Hole. One extending from Oil City up the western bank of the Allegheny to the mouth of Pit Hole creek, to which it is proposed bringing the oil by means of pipes; the other extends from Reno, accross Oil creek, to Cherry run, thence up Cherry run to Plumer. Both these roads connect with the Atlantic and Great Western Railroad.

There are a number of fine wells near the mouth of Pit Hole creek and above, along the Henry and McCrea farms. These wells are from three to six hundred feet deep. Some of them have flowed, and are now pumping from five to twenty barrels per day.

As an indication of the strata, the following register is given of a well bored on the Culbertson farm :—

| | | |
|---|---|---|
| To the rock | 33 | feet. |
| Thickness of first shale | 127 | " |
| " " sand rock | 25 | " |
| " second shale | 60 | " |
| " " sand rock | 40 | " |
| " third shale | 22 | " |
| " " sand rock | 28 | " |
| " fourth shale | 215 | " |
| " in fourth sand rock | 50 | " |
| Total | 600 | " |

In the second rock there was a rose tint, approaching a flesh color, that has frequently been perceptible in these deep sand rocks. One of the wells on the Hussey & McBride farm flowed from two to two hundred and fifty barrels per day for two months; it was then closed up as there was little demand for oil. It has since been pumped with an increase of yield; it now produces some fifteen barrels daily.

A little farther up, Hemlock creek empties upon the eastern side of the river. This region is sometimes known as President, that being the name of the township and post-office. This creek, in connection with the river, presents a large extent of flat lands for the consideration of parties that prefer low lands and proximity to water courses as locations for wells. These lands are now in process of being tested, and the successes that have attended operations below would indicate that Hemlock and President will not be lacking when vigorous operations shall have explored their territory.

As we approach the Tionesta, which is thirty miles above Franklin, numerous islands dot the river, some of them of considerable extent, that are sought after as oil territory. Formerly islands were in better repute than at present. It was supposed that the water courses indicated the course of the oil currents, and that under the bed of the river was the best possible location. Theory and observation both seem to agree latterly that this is not the case; that there are no oil currents; that oil wherever it may be found is sluggish; and that it may be found underneath the high lands as well as on the water-courses. Still these islands are eagerly sought after, and in many cases prove very productive.

Tionesta is a beautiful town on the eastern bank of the river, and was, at one time, the centre of a very thriving lumber trade. Of late this has somewhat fallen off in importance, as the people have been giving their attention to other pursuits. Little Tionesta empties below the town, while Tionesta proper enters the river at the town, and, with its tributaries Coon creek, Ross and Salmon runs, affords a large extent of available land for experiment and development. These operations are meeting with success, too, in proportion as they are prosecuted with vigor and perseverance.

Four miles above Tionesta we come to West and Little Hickory creeks, one upon each side of the river, and three miles farther up East Hickory. Operations were commenced on these streams early in the history of the oil business; but were interrupted by the flowing wells on Oil creek, together with the circumstances of the war; still there are some wells in operation, and preparations for great and enlarged operations are in progress by stock companies. The geological features of this region have a great resemblance to those of Oil creek, and future operations may be attended by like success.

A short distance above this is Tideoute, once the scene of extensive and exciting oil operations, bidding fair to rival Oil creek; of late there has not been so much done, as circumstances have interrupted operations to a considerable degree. This place is in Warren county.

We return now to the lower Allegheny, or that portion of the river below Franklin. There is here a large extent of valuable land for oil purposes, and, so far as matters have progressed, operations have been attended with the most gratifying success. Just below Franklin there are quite a number of successful wells, or those yielding oil. The comparative success attending operations on this portion of the Allegheny is greater than in any other portion of the oil region. On an average more than half the wells bored have proved successful. In a distance of two miles below the borough line of Franklin, there are at present, on the western bank of the river, thirty wells bored; of these, eighteen yield oil in remunerating quantities, and six, or one-third, are or have been flowing wells. The aggregate is now about one hundred and fifty barrels daily, having a gravity of from 34° to 37°, Baume. In this region the depth of the wells is about four hundred and fifty feet, and the oil is found in the second sand rock. The variation in depth and in strata is so small as to be unworthy of note.

For the same distance, on the eastern side of the river, the number of wells is about forty-two. The ratio and amount of productiveness is probably about the same as that on the western shore. A fine well opposite Franklin is now yielding some fifty or sixty barrels daily.

It will be observed here that all the wells as compared with those of the upper Allegheny, and particularly those of Oil creek, are shallow, not having advanced below the second sand-rock. When operations are carried farther below, to the third and fourth sand rocks, new and better developments may be expected. It was on this part of the river that the first well below Franklin was opened, known as the Hoover well. The depth was three hundred feet, and produced, by pumping, forty barrels daily, for perhaps two years, when the proprietors thinking to improve it by boring deeper, got the boring implements fast beyond remedy, and the well has been a sealed fountain ever since.

Five miles below Franklin upon the eastern bank is East Sandy creek. Good success has attended operations here. A flowing well, yielding one hundred and fifty barrels per day, has given an impetus to operations that will go far towards bringing to light the hidden resources of this region.

Four miles farther down is Big Sandy creek, a noble stream that branches into numerous divisions. Its principal branches are Little Sandy and South Sandy. Upon this stream is Waterloo, about five miles from Franklin. Frowning hills, whose sides are covered with boulders, overlook this stream. Many portions of the hills present a broken and shattered appearance as though by some vast force they had been rent asunder to afford a channel for the creek. If these appearances indicate anything, this region should be a favorite place of resort for oil men. This has been the case, and the banks on both sides of the stream have been purchased for many miles from the river.

There are some islands in this part of the river, that have been explored with great success. There is one practical inconvenience connected with boring on islands; it is difficult to get supplies of coal to the wells, and often impossible to ship the oil when produced. In addition to this, they are subject to inundation at times of great freshets. Some of these floods are terrible in their power and devastation, particularly at the breaking up of the ice in the spring, bearing all before them, and sweeping away every thing that comes within the sphere of their influence. Still island lands are sought after with great eagerness, and purchased at extraordinary prices, with a disposition to risk all accidents from the spring flood, and to endure all inconveniences from the low water of summer.

More attention has been attracted to the lower Allegheny at the present time than ever before. Large amounts of lands have been purchased, and extensive companies, with capital sufficient to explore thoroughly, have already commenced operations, and ere long magnificent results may follow. It is an interesting fact, too, that with all the success that has attended operations here, not one well has yet reached the stratum that yields oil the most liberally on Oil creek. From the dip of the strata, this would be in this region from one thousand to twelve hundred feet. When some bold operator has the courage to disregard surface shows, and faith to push his way downwards, it may be that a nearer approach to the oil caverns will result in a more copious supply than has yet been realized.

From the general direction of the streams along which oil has thus far been chiefly found, the opinion has been entertained that some immense deposit has been accumulated at the point where all these stores converge. The general direction of the river above and below, of East Sandy, French creek, Sugar creek, Two-mile run, Oil creek, Tionesta, and the Hickories, would cause them all to converge at a point not far from Franklin, and in that locality we might look for the great oil basin. The indications are, certainly, that this is becoming more and more the central point of the producing region, as new developments down the river and up French creek are enlarging the area; but there is little or no reason to suppose that the oil is found tending in any direction. It seems evidently sluggish, and confined to particular cavities, throwing out veins and crevices in various directions, which veins are generally filled with water. Still there may be some vast receptacle or lake of oil somewhere in the oil region, connected with other lakes of smaller dimensions, and these connected with small caverns or ponds, and the whole connected by a series of veins ramifying in endless profusion throughout all this region.

The region along the Allegheny is not without its traditions. Some of the aged people here remember an old Moncey chief named Ross. The old brave always asserted that there were silver mines along the Allegheny. At one time he proposed pointing out one of these mines to an old citizen, then of Franklin. It was said to be situated in a ravine between Franklin and Oil City. After leading the white man up this ravine, where umbrageous trees and moss-covered rocks made a gloomy and fearful shade, they came to a second ravine, cutting the first upon the right, where ragged rocks and irregular banks suggested the work of an earthquake, and passing up the second for a short distance the chief suddenly paused, and with solemn emphasis said: "I dare not go farther. The mine is within five rods of you; find it for yourself."

---

## CHAPTER XVI.

### USES OF PETROLEUM.

Here a large field opens before us. It has not yet been fully explored. This branch of the subject is yet in its infancy. The great miracle of the nineteenth century, although grasped practically in many of its details, has not yet been submitted to that nicety of treatment and delicacy of manipulation that will bring its hidden uses to full view, and render every portion of it an element of value, tending to the use, and comfort, and mental elevation of mankind. It is now made to pass through the fire, and many useful and beautiful products have been the result; yet much of that which is, no doubt, valuable, goes to comparative waste, and is reckoned as loss. Hereafter, in the chemical laboratory, under the treatment of an eager and inquisitive science, it will, no doubt, yield rich and beautiful results, that will convey to the world still stronger evidences of the goodness of the great Benefactor in the gift of this wondrous wealth.

It has long been used in its natural state for many useful purposes, but not until recently has it been made to pass through a course of treatment to adapt it to a variety of useful purposes. This course of treatment has not yet attained to perfection. In the refinery for illuminating purposes, in its treatment for lubricating purposes, and, above all, in the laboratory treatment for the extraction of its finer and more delicate properties, it is, as yet, almost a terra incognita.

As an illuminator it stands pre-eminent in the form of refined oil for the lamp. The days of tallow candles have passed away forever. Their reign has been terminated, and a better dispensation has taken their place. Long years ago the early settlers used it in a crude state, in rude lamps, but it was not a satisfactory agent. The openness of the early log cabins were rather favorable to its use, inasmuch as the smoke and vapor that accompanied its use could readily pass off; but it was never used save in cases where no other light could be obtained.

Even after the refining and deodorizing process had been resorted to, the imperfection of the lamps then in use rendered the refined oil a very imperfect article for giving light. It was then used in lamps furnished with cylindrical tubes, about an inch and a half in length, and without chimneys. The light was but feeble, and liable to be extinguished with even a slight breath of air, and the lamp could not be carried from place to place. Of necessity, the oil was then expensive, as it was found only in certain localities and in limited quantities, and, under these circumstances, it soon went out of use.

When the discovery of the distillation of oil from cannel coal brought this fluid into use, improvements were made in lamps so that it could be used not only with economy, but with pleasure; and when the large wells of petroleum were opened up, the improved lamps brought it into immediate use. After distillation and deodorizing, by processes discussed in a future chapter, it stands chief, in many respects, amongst the illuminating substances. Its use is characterized by a clear, strong and steady flame, that is most agreeable to the eye, and so different from the staring, flickering light afforded by a jet of gas.

As an illuminator it is used with perfect safety, where even ordinary care is exercised. It is important, however, that care and precaution should be used both by the refiner and the consumer. If the benzine is not well removed it is an explosive mixture, and cannot be used with safety. And, on the part of the consumer, ordinary care should be used to guard against accidents. The safety or danger of the oil may be tested by pouring a small portion in a spoon and applying a match. If it takes fire readily it betrays the presence of benzine, and is unsafe; if it will not ignite, it may be considered safe; but lamps should not be filled near the fire, nor when already lighted.

Neither the illuminating qualities, nor the economy of the light produced by petroleum, have as yet been properly appreciated. By way of contrast with other illuminating substances, a table, prepared by Professor J. B. Wethee, of Ohio, is here presented:

| Articles Used. | Intensity of light. | Quantity of Light from an equal measure of oil. | Cost of an equal quantity of light |
|---|---|---|---|
| Coal Oil Petroleum... | 13.70 | 2.60 | 4.00 |
| Camphene ............... | 5.00 | 1.30 | 4.95 |
| Whale Oil............... | 2.40 | 85 | 12.00 |
| Lard Oil................. | 1.50 | 74 | 17.00 |
| Sperm Oil............... | 2.00 | 95 | 26.40 |
| Burning Fluid......... | 75 | 30. | 20.34 |

From this table it will be seen that refined petroleum is superior to all other oils and burning fluids, both in intensity of light and in cheapness of supply. It is rapidly finding its way into foreign countries, and interfering sadly with the olive orchards of Italy and other portions of southern Europe, as it will, ultimately, with those of Asia. It would seem that through its influence one of the most ancient and even sacred productions of the East is to fall into desuetude, and the vineyard and the olive yard cease to be enumerated together, in speaking of a fertile and prosperous land. The olive seems to be destined to pass away, with all its sacred and hallowed associations, in the East, just as the time-honored tallow candle, with its associations of home, and fireside, and family circle, has already passed away from our midst, to come not back again.

As a lubricator, in various forms, petroleum stands pre-eminent. Sometimes it is used in its crude state, and sometimes with certain preparation, that is discussed in a succeeding chapter. It is found adapted alike to the roughest and heaviest machinery, and to the smallest and finest, and is, at the same time, most economical and easily applied. In combination with other substances it can be made to retain its place, where other oils soon wear out.

As a medical agent it has long held a prominent place in the estimation of the world. It is quoted in the oldest Hindoo medical authorities as occupying a prominent place in the estimation of the medical faculty of that day, and was well known from that day to this, as it is still obtained largely in the Hindoo country. It was well known to the Indians when this Continent was discovered by Europeans. Its use is alluded to in many of the earlier histories of this country.

As to the estimate placed upon it at a very early day, it is interesting to quote from an article published in the "United States Magazine," for July, 1792. "In the northern parts of Pennsylvania there is a creek called Oil creek, which empties itself into the Allegheny river, issuing from a spring, on the top of which floats an oil similar to what is called Barbadoes Tar, and from which may be collected by one man several gallons in a day. The American troops in marching that way halted at the spring, collected the oil, and bathed their joints with it. This gave them great relief, and freed them from the rheumatic complaints with which many of them were afflicted. The troops drank freely of the waters; they operated as a gentle purge."

We have another ancient authority, that if not particularly valuable is curious and rare. The quotation is from an old work published in London nearly a century ago. Its title in part is:—"History of the Missions of the United Brethren among the Indians of North America. In Three Parts. By Henry George Laskiel. Printed for the British Society for the Furtherance of the Gospel. 1794."

"One of the most favorite medicines used by the Indians is the Fossil Oil (Petroleum) exuding from the earth, commonly with water. It is said that an Indian in the small pox laid down in a morass to cool himself, and soon recovered. This led to the discovery of an oil spring in the morass, and since that time many others have been found in the country of the Delawares and Iroquois. They are observed both in the running and standing water. In the latter the oil swims on the surface, and is easily skimmed off; but in rivers it is carried away by the stream. Two have been discovered by the Missourians in the river Ohio.

"This oil is of a brown color, and smells something like tar. When the Indians collect it from standing water, they first throw away that which floats on the top, as it smells stronger than that below it. Then they agitate the water violently with a stick, the quantity of the oil increases with the motion of the water, and, after it is settled again, the oil is skimmed off into kettles, and completely separated from the water by boiling. They use it chiefly in external complaints, especially in the headache, toothache, swellings, rheumatism, dislocations, &c., rubbing the part affected with it. Some take it inwardly, and it has not been found to do harm. It will burn in a lamp. The Indians sometimes sell it to the white people at four guineas a quart."

Allusion has already been made to its use by the early white settlers in this region. Oftentimes far from regular medical advice and assistance they learned to treat themselves for all minor complaints and difficulties, and often with great success. For external injuries, such as cuts, bruises, and dislocations, they found it an almost sovereign remedy, and thought they could send their friends in the East no more acceptable present than a bottle of Seneca Oil for medicinal purposes.

In very many instances it has proved valuable as a disinfectant. An instance is recorded where an epidemic that was raging in a town in Italy was stayed and ultimately driven away by a liberal use of it on the streets and hospitals. There is no doubt but that advantages would arise from a free use of it in the hospitals of our own country, ministering to the comfort of patients and nurses, if not neutralizing the noxious odors that abound.

When incorporated with soap, it is said to act beneficially upon the person in various cases of cutaneous affections. Petroleum soap has already found its way into market, and although as yet used medicinally, will, no doubt, come into use as an important article of luxury in toilet arrangements, adding to present comfort, and preventing cutaneous affections that would be otherwise unavoidable.

The vapor of petroleum has been used to much advantage in cases of weak and diseased lungs, as well as in asthma. It is taken by inhalation. For the same class of diseases it has been taken internally in some cases with quite marked success. But this matter is left to the disposition of the medical faculty, who will, no doubt, give to it all due attention.

The effect of petroleum upon operatives, and others brought in constant contact with it, is certainly not deleterious. There is no evidence that it acts in any case in a manner unfavorable to health; but, on the contrary, there is evidence that it operates beneficially in promoting strength, and increasing the flesh of those who are in the constant habit of handling it. There is one curious fact that has been observed in relation to workmen engaged at the well, and that is, that their hair and beards soon become luxuriant, dark colored, and assume every appearance of vigorous health and growth.

The testimony of an assistant-surgeon from the bloody field of Gettysburg, will be here in place, as to the effects of petroleum in the medication of wounds:—"What water is to a wound in an inflamed state, petroleum is to a wound in a suppurating state. It dispels flies, expels vermin, sweetens the wound, and promotes a healthy granulation. Having dressed the wounds of two patients with it, I found they had sunk to sleep before I had finished the dressing of the third."

Dr. Ducaisne, a physician of Antwerp, has given his opinion in regard to the medical properties of this substance, in stating that by fumigating the apparel of persons exposed to epidemic disease with the gas arising from it, it will afford protection from attack; also, that it may be applied successfully to those cutaneous diseases that are engendered by the presence of parasitical animalcules.

As an article of fuel, petroleum is no doubt destined to occupy an important place. Sufficient time has not yet elapsed since its production in large quantities to carry any important experiments to final success; but the very nature of the material would suggest that it might be profitably used where economy of bulk and weight was particularly desirable, as on board steamships. It is simply concentrated fuel. And if any apparatus can be invented in which concentrated fuel can be used safely and profitably the problem will be solved. The grand difficulty attending ocean steam navigation now is that the coal necessary to complete the voyage is nearly sufficient to freight the ship without anything further.

Now petroleum can be safely used. By a very simple process, keeping it for a time up to 212° Fahrenheit, by the admission of steam pipes to the tank it can be deprived of the greater part of its explosive matter, and rendered practically safe. Even the 29° oil of French creek will not take fire when a lighted match is applied. As to the apparatus to be used, anything that would disperse it, and bring it regularly and gradually to the fire surface would be all that would be required.

Some experiments have already been made by the United States Navy as to the practicability of bringing it into use in our own ships of war; but no report has yet been made public in regard to the matter. The matter has also been discussed in England. A paper was recently read there on the use of mineral oils as fuel for steamships, in which it was asserted that twenty gallons of oil were equal to one ton of coal, and that the heating power of the oil was to the coal as four or four and a half to one. This is a low estimate in favor of the coal, as one ton of cannel coal will produce forty gallons of oil; but if we assume forty gallons of oil, or one barrel, as the equivalent of one ton of coal, we will see the vast advantage the oil has over coal both in economy of weight and space. A barrel of oil weighs probably three hundred pounds against twenty hundred of coal, or nearly one-seventh, and as to bulk the proportion is nearly as great.

If proper machinery can be produced for its use there is no reason why a ship might not carry sufficient fuel to make the tour of the world; or for ordinary voyages, the space now occupied by coal could be used for other freight, and four hands would be sufficient to man the ship, and conduct its operations.

This product is admirably adapted to the manufacture of gas. It possesses all the properties of coal that are necessary to this object, and has some advantages on the score of freight where it is necessary to transport it to a distance. As yet the crude oil is not much used for this purpose, as a more profitable disposition can be made by passing it through the distilling process, and using only certain portions for the production of gas. In the hands of some distillers there is a portion of tar produced from the residuum that is of great value in the manufacture of gas. It has been used with complete success in many of the factories of New England, and machinery has been invented and patented adapted to its use.

There is another product of distillation that is used for the same purpose. It is the first product after the operation of distilling commences, called gassoline, very volatile and explosive, but yet adapted to the purpose by a simple kind of machinery. There is a slight defect in the machinery used for making gas that may yet be obviated: it is the tendency of the gas to condense when the temperature is below freezing point. If this is obviated then the way is opening up for lighting isolated houses and churches with gas where there can be no regularly organized gas companies.

Paraffine is prepared from oil in small quantites. This is "a tasteless, inodorous, fatty matter, fusible at 112°, and resisting the action of acids and alkilies. It is so named from its little affinity for other substances."—*Brande.* This substance seems to be a product of distillation rather than of mechanical separation. More paraffine is separated when the distillation is carried on at a high temperature than at a low one, as is indicated by the color of the distillate. When separated from the lubricating oil it is placed in bags and pressed, after the manner of pressing linseed oil. It then appears as a brown mass, and, after being washed in naptha, in which it is not soluble, the pressing is resumed until it assumes the hard white substance that is seen in the shops. It is used chiefly in the manufacture of candles. It is also used in the manufacture of chewing gum, but, after all, is not as valuable a property as some others found in the oil.

The naptha, or benzine, a substance that results from distillation, and makes its appearance from the condenser previous to that of the burning fluid, was at one time considered valueless, and thrown away. It is now considered valuable for many purposes. In the various uses to which turpentine was formerly applied, this substance forms a good substitute. In paints and varnishes it answers a good purpose, though, perhaps, not equal to turpentine. It has this good property, however: it is an admirable drier in paints and varnishes, where it is used, drying and forming a body with great rapidity. The same substance, when nicely deodorized, is used with great success as a renovator, removing paint and grease as though by magic, and leaving no trace behind. It can also be successfully used as a medium for the preparation of perfumery, as it can be thoroughly deodorized, and, when so prepared, does not differ greatly from alcohol. It is even asserted that brandy has been extracted from it by chemical processes. This is not at all unlikely, and yet it is not probable that the cause of temperance will ever bring any serious charges against the petroleum trade on the ground that it contributes to the manufacture of ardent spirits.

The heavier qualities of oil have been used successfully in the preparation of leather. Some tanners doubt whether it has sufficient body to make good leather, but a little preparation would, probably, obviate this difficulty. The heavy substance manufactured for lubricating oil would, probably, be found to have sufficient body for all practical purposes. In the manufacture of patent leather, the heavy oil of French creek and Franklin is said to answer most admirably. One manufacturer asserts that it answers the purpose better than any substance he has heretofore employed.

But the aesthetics connected with the distillation of this oil must not be passed by in silence. On a pleasant, sunshiny day we see a bright globule of petroleum rising from the bed of the stream. It has had a long journey in reaching the regions of upper air from its bed deep down in the rock. It has forced its way through the crevices until, at last, it emerges from the bottom of the stream. As it reaches the surface of the water it disperses, and, as it glides away, all the colors of the rainbow are reflected from its undulating surface.

"What radiant changes strike th' astonished sight!
What glowing hues of mingled shade and light!
Not equal beauties gild the lucid west
With parting beams o'er all profusely drest.
Not lovelier colors paint the vernal dawn,
When Orient dews empearl th' enameled lawn,
Than in its waves in bright suffusion flow,
That now with gold empyrial seem to glow;
Now in pellucid sapphires meet the view,
And emulate the soft celestial hue;
Now beams a flaming crimson on the eye,
And now assumes the purple's deeper dye.
But here description clouds each shining ray—
What terms of art can Nature's powers display."

That thin pellicle on the water's bosom is lighter and more evanescent than the morning mists, yet it possesses a world of beauty. We gaze upon those colors, ever changing in their lustre and variety, until imagination revels in its most delightful dreams, suggesting thoughts of the good and beautiful, and reminding us how beauty lingers amid the most unpromising things of earth! Just as the bow that spans the mantling cloud reminds us of all beautiful things that glow around its antitype which spans the emerald throne on high, so, as we gaze upon the prismatic tints that are reflected from the glowing surface of the oil globule, we dream of all that is beautiful in colors and gorgeous in tinted radiance as being hidden amid the elements of petroleum.

This dream has its fulfilment amid the processes of distillation and treatment. One product in this treatment is called Aniline; that is the base of those beautiful colors in which ladies so greatly rejoice—such as Mauve, Magenta, Solferino. There is also a shade of blue called Cerulean, that causes one to dream of the calm blue sky after the storm has subsided; and still another, called Azurine, soft and delicate as the azure vaults above. And in process of time, no doubt, the most delicate colors for flower and landscape painting will be educed, that will give a new impetus to the fine arts and to the development of taste in our midst. But the whole matter relating to petroleum is yet in its infancy, and wonderful results yet await development in the processes of experiment and treatment.

# CHAPTER XVII.

## REFINING.

THUS far in the history of petroleum the most important use to which it has been applied is that of an illuminator. Without it the world would now seem going back to darkness. It is particularly valuable in the country, and in small towns, where the population cannot be supplied with gas. And even where gas companies are in operation it is valuable as affording a soft, steady light, more grateful to the eye than that of gas.

The oil as found in the wells is generally unfit for burning in lamps. It was formerly used in the open air, in a rude kind of lamp or lantern, but with little satisfaction, as the smoke and odor accompanying the combustion of the crude oil were absolutely intolerable. But, by the refining process, aided by lamps of suitable construction, these difficulties are all removed, and, as far as an illuminator is concerned, petroleum is the great blessing of the nineteenth century.

The matter of refining is simple. It consists in removing the coarse, heavy ingredients that partake much of the nature of coal, and preserving the lighter portions.

In this process, too, the volatile, explosive fluid is separated, and the offensive smell almost entirely removed. This process was somewhat understood many years ago, and brought into requisition to prepare the small quantities of oil obtained by ditching and collecting with blankets. This, of course, was on a very limited scale, and not brought to much perfection. It was also brought to bear in the finishing of coal oil, obtained by distilling cannel coal. But there was no great stimulus to experiment or research. It was more curious than profitable. Modern oil operators have applied all the stimulus that was necessary. Enterprise, and science, and energy have been brought to bear in the matter, and now the operation of refining and preparing oil for illuminating purposes has been brought to a very high state of perfection.

This matter was at first commenced on a very small scale in the oil region. At one time in Franklin, through the energy of a physician and an extemporized still, oil was manufactured each day for the wants of the town as it was required; and it was an honest source of pride to feel that the oil was produced and refined within the limits of the town where it was consumed.

But the matter soon began to enlarge. Regular apparatus was provided, and regular establishments erected, until the business of refining oil has become a prominent and important one, not only in the oil region, but in distant places. It seemed at first natural that the business of refining should be carried on near where the oil is produced, by way of economizing freight, but latterly almost every portion of the oil is utilized, so that there is little or nothing lost in the matter of freight, and frequently there is an advantage in refining in places where coal and labor is not so expensive as they are in the oil valleys.

The business of refining, however, is largely carried on

in the oil region. In Venango county alone there are about ninety refineries, with a capital of about a million dollars, or an average of a little more than ten thousand dollars each. This business is carried on with success in the process, and profit in the final result, although, doubtless, there is still great room for improvement both in the appliances and in the manipulation.

The products of the oil in the process of distillation are much the same in every case, yet they differ much in kind and quantity. Sometimes the process is not carried out to the utmost capacity of the oil in producing burning fluid, and a larger proportion turned in other directions. Sometimes it is considered most profitable, and so most desirable, to reduce almost the entire mass to the one product of burning oil. The views of distillers differ in this matter, and the location of the refinery has its influence in determining the process. In regions where fuel is scarce and high, the residuum is profitable for fuel. In near proximity to coal fields it is better to use it in a different direction.

The usual products of the process are benzine or naptha, burning fluid, lubricating matter, paraffine, and, when carried to its utmost extent, coke. Paraffine is not found in large quantities, and in the smaller refineries is not regarded as of sufficient importance to warrant particular attention. In addition to these, there are some resulting substances that may be prepared after the deodorizing process has taken place, such as copperas and phosphate of lime, that are sometimes made profitable, particularly in connection with large refineries.

The business is carried on in works of various capacities. The stills seldom contain less than ten nor more than one thousand barrels; usually the capacity is from twenty to fifty barrels. The stills are of wrought iron, set in masonry, with a furnace underneath for the adjustment of fire. The still is simply a mighty retort, with its pipe connected with a condenser, or long coil of pipe, in a vessel of cold water, in which the vapor, driven off by heat, applied to the retort is condensed into benzine and "distillate," as the new product is termed.

The chemicals used are sulphuric acid and caustic soda. These are employed in relieving the distillate of the intolerable odor that clings to the oil in almost all stages of the manufacture. Ordinarily, the smell of petroleum, when freshly drawn from the well, is neither injurious nor offensive—it is even pleasant; but in the process of refining and deodorizing it is as bad as imagination can well conceive. There are as many distinct smells traceable around a large refinery as Coleridge alleged could be detected at Cologne. But, withal, the process of refining does not seem to work unfavorably to the health of the workmen.

In the process of refining, as now pursued, the still should be filled with crude oil to within one tenth of its capacity, the man-head securely packed, and the fire kindled underneath. As the mass within rises in temperature, a portion of the water that is found in all crude oil will separate and be drawn out at the proper orifice. It is remarkable with what tenacity the salt water and oil cling together. They have sojourned in loving companionship so long that it is hard to separate them. The union is like that of the prairie dog and owl, and even snake, in the same burrow in the western prairies—dissimilar in nature, and yet bound together by a community of interest.

As the mass becomes warm, the first product of the

condenser is a very light, volatile substance, called Gasoline, of a specific gravity of about 80° Baume. After this, and when the oil has reached a temperature of 212° Fahrenheit, the remainder of the water in the still will pass off in the form of steam, and be condensed as acid water. The product of the condenser is then benzine or naptha, from 80° down to 60°, when burning fluid begins to present itself; from 60° down to 38° the product is different grades of burning fluid, that are usually mixed, forming a light-colored oil when finished. Burning fluid, for economical and safe use in the household, should not be above a specific gravity of 46° Baume, and endure a "fire test" of 115° Fahrenheit, and not flash in the lamp when burning.

The product of the still under 38° is treated differently at different establishments. Sometimes it is distilled, and a further per centage of burning fluid extracted; sometimes it is used for lubricating purposes, and at still other establishments the paraffine is extracted.

The practice differs materially in different parts of the country, both as affected by the views of different distillers, and the comparative value of different products in the process. The practice is affected, too, somewhat by the character of the crude oil used. This differs in specific gravity from 38° to 50° Baume. The heavier oils, however, are seldom or never distilled, being more valuable for lubricating purposes. In some establishments the oil of the Allegheny and that of Oil creek are mixed, and in this way a compound of medium gravity is obtained that works most satisfactorily.

The gasoline and naptha are then turned over to be treated, by being steamed down to a proper degree of gravity, for other manufacturing purposes. The distillate is now ready for the deodorizing process. It is placed in a treating tank of iron or wood, lined with lead, called an "agitator," where it is agitated with a paddle-wheel, or, what is better, with a powerful wind blast, produced by the power of a steam engine. When this agitation resembles a miniature storm on the ocean, a portion of sulphuric acid, amounting to from three-fourths of one per centum to one and a half per centum, is added slowly, and the mass agitated from ten to thirty minutes, according to the views of the chemist. When the tank becomes quiet, and the mass settles, the impurities are drawn off from below, and one or more treatments made in like manner.

When sufficiently treated with acid, and the impurities removed, the treatment is resumed in the same manner with soda lye, or caustic soda dissolved in water. After being thoroughly agitated with this addition, and the further impurities withdrawn, lukewarm water is freely sprinkled upon the oil and allowed to run off, until the oil is found to be entirely free from impurities, when it is ready for market.

The acid cannot be again used in this process, but need not be lost. It may be digested with scraps of iron, and afterwards roasted and made into copperas. The alkali may be used again and again, until it parts with its caustic properties. This may be restored by the addition of quick lime, and finally may be made to act upon bones, producing phosphate of lime.

By way of contrast with the modern results of the refining process, the following extract of Professor Silliman's report, made in 1855, is appended:—

"Three hundred and four grammes of crude oil,

submitted to fractional distillation, in a linseed oil bath, gave:—

| | | Temperature. | | Quantity. |
|---|---|---|---|---|
| 1st | product, | at 100° C. = 212° F. | (Acid water.) | 5 grms. |
| 2d | " | at 140° C. to 150° C. = 284° to 302° F. | | 26 " |
| 3d | " | at 150° C. to 160° C. = 302° to 320° F. | | 29 " |
| 4th | " | at 160° C. to 170° C. = 320° to 338° F. | | 38 " |
| 5th | " | at 170° C. to 180° C. = 338° to 356° F, | | 17 " |
| 6th | " | at 180° C. to 200° C. = 356° to 392° F. | | 16 " |
| 7th | " | at 200° C. to 220° C. = 392° to 428° F. | | 17 " |
| 8th | " | at 220° C. to 270° C. = 428° to 518° F. | | 12 " |

| | |
|---|---|
| Whole quantity distilled | 160 |
| Leaving residue in retort | 144 |
| Original quantity | 304 |

"*Product* No. 1, almost entirely water.

"*Product* No. 2, an oil perfectly colorless, very thin and limpid, but having an exceedingly persistent odor, &c.

"*Product* No. 3, tinged slightly yellow, perfectly transparent, &c.

"*Product* No. 4, more decidedly yellowish than the last, &c.

"*Product* No. 5, more highly colored, thicker in consistency, and had a decided empyreumatic odor.

"*Product* No. 6. This with the two subsequent products, were each more highly colored and denser than the preceding. The last product had the color and consistency of honey, and the odor was less penetrating than that of the preceding oils.

"The *density* of the several products of this distillation shows a progressive increase, thus:

| | | |
|---|---|---|
| No. 2 | density, | .733 |
| No. 3 | " | .752 |
| No. 4 | " | .766 |
| No. 5 | " | .776 |
| No. 6 | " | .800 |
| No. 7 | " | .848 |
| No. 8 | " | .854 |

"To form an idea of the density of these several products, it might be well to state that sulphuric ether, which is one of the lightest fluids known, has a density of .736, and alcohol, when absolutely pure. .815."

In regard to the properties of this oil, Professor Silliman says:—"Exposed to the severest cold of the past winter, all the oils obtained in this distillation remained fluid. * * * The chemical examination of these oils showed that they were all composed of carbon and hydrogen, and probably have these elements in the same numerical relation. * * * The oils contain no oxygen, as is clearly shown by the fact that clean potassium remains bright in them. Strong *sulphuric acid* decomposes and destroys the oil entirely; *nitric acid* changes it to a yellow, oily fluid, similar to the changes produced by nitric acid on other oils; *hydro-chloric*, *chromic* and *acetic acids* do not affect it; *litharge* and other metalic acids do not change it or convert it in any degree to a drying oil; *potassium* remains in it unaffected, even at a high temperature; *hydrate of potash*, *soda* and *lime* are also without action upon it.

"The oil is nearly insoluble in pure alcohol, not more than four or five per centum being dissolved by this agent. In ether the oil dissolves completely, and, on gentle heating, is left unchanged by the evaporization of the ether. India rubber is dissolved by the distilled oil to a pasty mass, forming a thick black fluid, which, after a short time, deposits the india rubber. It dissolves a little amber, but only sufficient to color the oil red. It also dissolves a small portion of copal in its natural state; but, after roasting, the copal dissolves in it as it does in other oils."

Thus far Professor Silliman; but it must be remembered that the oil submitted to his examination was the thick, heavy substance collected by ditching and forming pits. The results would have been very different with the lighter oils of Oil creek, brought up from the depths below. Still the report is valuable, as showing the properties of the oil under the most unfavorable circumstances attending the examination.

---

## CHAPTER XVIII.

### LUBRICATORS.

We come now to speak of petroleum as a lubricator. An unexceptional article of this kind was becoming one of the great wants of the age. With the multiplication of machinery, and the increase of business throughout the land and on the ocean, the want of an article adapted to all kinds of weather was exciting the inquiry of various classes of business men connected with railroads and manufactories everywhere.

Petroleum appears to answer the conditions of such an article better than any other substance yet discovered. It has natural characteristics that adapt it to the purpose, is found in exhaustless quantities, and is produced at comparatively low rates.

It has already been remarked that this oil is found with various degrees of densities. At Franklin, on French creek and Sugar creek it is sometimes found as great as 29°. This heavy oil is used almost exclusively for lubricating purposes, being much better adapted for the purpose than the lighter and more volative oils of Oil creek. There does not seem to be any difference in these different oils, save that the heavy oils being found nearer the surface, have parted with their naptha to a much greater extent than those found deeper in the rock. These heavy oils are used for various kinds of machinery without any preparation whatever, and for heavy rough machinery answers a very good purpose; for finer and more delicate works some little preparation seems to be necessary.

For a year or two after the opening of the flowing wells, great quantities of oil was found floating upon the surface of the river. The great wells that were yielding many hundred barrels per day without tankage, and the frequent wrecking of boats during pond freshets in Oil creek, were attended by very great waste. The result was that for miles down the Allegheny, oil could be collected in almost unlimited quantities. The mode of procedure was to throw a boom out from the eddies along the shore at an angle of about forty-five degrees. This boom received the oil while the water passed beneath. Sometimes instead of a boom, an old flatboat was thrown out in the same manner, and the oil thrown into the boat with dippers and buckets. On the shore an arrangement was provided for boiling, or rather heating, the oil, in order to draw off the more volatile portions, as well as separating any earthy matter that might be mingled with it. It was also strained to remove impurities. In this way the oil is fitted for lubricating purposes. In its voyage on the surface of the water it has parted with a large portion of its naptha, but has taken up many impurities. By the heating and straining these are nearly removed, after which it is placed in barrels for market.

At one time there was an extemporized establishment of this kind for every half mile of the distance from Oil City to Franklin, and even below for some distance. At some of these points were gathered at times ten barrels per day, being equal to a moderately sized oil well.

At some of the refineries lubricating oil is prepared in large quantities. The practice differs at different establishments. At some only about fifty per cent. of illuminating fluid is run off from the stills, and the remainder prepared for this purpose. At others, there is a much larger proportion of illuminating matter extracted, while at other stills very little lubricating oil will be produced. This will depend somewhat on different causes, gravity of oil, value of lubricator as compared with other products of the still, and different views of refiners. Experience and observation are leading to many changes in the manner of treating petroleum, and the indications now are that the lubricating oil will gradually cease in connection with refining, as all the residuum can be used to better advantage in other ways, and the manufacture be carried on independently with more profit, and perhaps yield an article of much better quality. Still all these processes are yet in their infancy, and must be governed by future developments.

There are difficulties in the use of this oil, however, even when treated in the manner described. There is a slight grit that still remains, manifesting itself in use by discoloring the journals of the machinery, showing that there is a constant but slight wear of the material. There is also a slight gum that forms in its use, that takes up and retains the dust, together with a slight trace of an acid that operates unfavorably upon the machinery, particularly that which is of a fine quality. Various experiments have been resorted to to set free these deleterious substances.

It is found on analyzing the petroleum that the grit is due to a species of rotten-stone that is held in solution with, and derived probably from its native beds. To this, its old companion in the rock, it clings most persistently, even after being brought to the surface, and it is only by a peculiar process that it can be separated from it. The slight trace of acid is also brought from the same place of repose, and held in equally close combination with it, until separated by chemical means. When these two operations are performed the petroleum begins to assume a very important place and value in manufactures.

Establishments are already in operation for the manufacture of this article, and its value and importance give the subject a growing interest. One of these, the "Great Northern Oil Company," have already commenced operations in Franklin with flattering prospects of success. Their principal article of manufacture is what is called the "Hendrick Lubricator," for which a patent is already secured in the United States, Canada, and England. This article appears to have been used with great success on our railroads, steamships, and in our manufactories for three years past, so much so that its success seems assured. Various grades of oil are manufactured adapted to different kinds of machinery and prepared by different processes.

For coarser and heavier machinery the oil is not distilled but refined. It is first placed in a large tank through which pipes filled with steam are passed keeping the oil up to a temperature of nearly 212° for a considerable length of time. By this process the inflammable portions of the oil are driven off, ensuring its safety, and at the same time not deteriorating its quality. The rotten-stone and mineral acid are then removed by a process known to the workmen, and a particular compound added that adapts it to the purposes in view, when it is deodorized, barreled, and ready for use.

For the finer and more delicate kinds of machinery a somewhat different process is resorted to. The oil is first distilled to remove the heavier and grosser portions. It then passes through the processes indicated above, and is adapted to any machinery from a sewing machine to a steam engine.

The characteristics of this lubricator appear to be valuable. It is entirely free from grit, gum, or acid, as is manifest from the appearance of the machinery where it is used, there being no discoloration of the journals by friction in their boxes. It does not become rancid by age as whale oil does; nor like lard oil does it thicken in winter and become thin in summer. There is another advantage this article in its manufactured state has over the crude. It is not inflammable nor explosive. While crude oil will ignite at 100°, this is entirely safe at 200°. Another valuable quality is that it does not congeal even when the mercury is at the freezing point, and can be used at all times with ease in the open air in winter.

This preparation has a strong affinity for itself, and manifests this by being tenacious and ropy to the touch, pre-eminently adapting it to the place it is designed to hold, at the same time it is diffusive and searching when applied to machinery. The finer qualities of this article differ in appearance from the best sperm oil only in the peculiar blue tinge that accompanies petroleum in all the varieties in which it is manufactured. This lubricator has the advantage of other substances of similar design, on the score of economy, being furnished at about half the price of sperm oil.

As a lubricator, this and similar articles manufactured from petroleum, are assuming a very great importance, and are, no doubt, destined to take the place of all other substances now in use.

---

## CHAPTER XIX.

### JOINT STOCK COMPANIES.

THE first companies that were organized for the development of the oil business did not succeed well. They were not incorporated as a general or even ordinary thing. There was very little responsibility connected with them, any further than as members were possessed of private property, and were convenient to the locality where operations were carried on. The members of these companies were held together by a rope of sand. They would withdraw at any time. The company was liable to be abandoned at any time, by discouragement, or failing means on the part of its members. When this took place, the property of the association was usually first applied to the liquidation of the debts, and sold for a trifle, and the balance of the indebtedness might be collected from the nearest member who had means within reach of the law.

This being the case these companies or associations became unpopular. Besides they were not calculated in the nature of things to carry out the purposes intended. There was really no capital raised. It was simply an assessment of such sums as were necessary to carry on the work from time to time, and, of course, the work was feebly done. Its only result must be disappointment and failure. In all the oil valley there is perhaps not a single case where one of these large, irresponsible companies carried forward their operations to the final advantage and benefit of its members.

No doubt the best possible way of carrying on the oil business is for one, or two, or three persons of capital and energy to procure a site, either by purchase or lease, and carry on the business under their own personal inspection. The prospects of success are greater in this way than in any other. But this cannot always be done. All men of energy have not capital, and all men of capital have not energy. Still capital must control labor, while the world remains as it is at present, and some plan must be adopted in order to give labor and small means an opportunity to engage in this business, even at the risk of injury from the capital.

In order to attain this end, and open the way for persons of small means to risk, if not benefit their available means, "joint stock companies" have been instituted and many of them carried on with great success and advantage to the stockholders. These companies are incorporated under the laws of the States in which they are organized, as there is an act of the Assembly of the State of Pennsylvania permitting "foreign incorporations" to hold lands and transact business on her soil. These laws differ in different States, and their application is different in different companies.

The manner of getting a joint stock company in operation is usually the following:—A piece of property supposed to be "oil territory" is acquired, or perhaps several pieces are acquired, by a few gentlemen who employ a solicitor to take the necessary steps to procure an act of incorporation. The titles to the property are examined, and a good and sufficient conveyance is executed from the owner to the corporation, in exchange for which the entire number of shares into which the company is divided is handed over to them. Sometimes there is a provision made for "working capital;" either so many shares are sold for this purpose, or the projectors place a given sum in the hands of the treasurer for this purpose.

This transaction takes place as soon as the company is organized and the officers appointed, and at the same meeting, a set of by-laws are adopted. The company is then in working order.

The shares have a certain par value, which must be plainly expressed on the face of the certificate, and on the books of the company. These shares may be sold, however, at a price below or above par as parties may agree; but whatever price the shareholder may pay for his shares he is not liable for any assessments thereupon, unless provision is made in the by-laws for such assessment. The directors or trustees are liable for the debts of the company if they do not set them forth in the annual statement required by law.

This is the general plan for the organization of these companies; but the principles and practice differ so much in different States, that it is impossible to go into all the details. The practice differs very much even in the same State. In some cases the whole stock is issued in payment for the land, in others a part only: the questions of capital, expenses, and other matters being arranged by each company to suit itself.

The number of these joint stock companies is something over six hundred, of which one hundred are paying dividends of from one to fifteen per cent. per month; but new ones are constantly organizing, particularly in the large cities. Of these six hundred companies, something over one-half of the number have their principal offices in Philadelphia; New York comes next in order, followed by Pittsburgh. Philadelphia has three hundred and twenty-nine; New York, one hundred and fifty-two; Pittsburgh, seventy-seven; Cleveland, nine; Boston, eight; Baltimore, eight; and Erie, four; the remainder are scattering. The nominal stock varies from fifty thousand to ten millions of dollars, the par value of the shares being from one to one hundred thousand dollars each. The nominal amount of capital represented by these companies is about three hundred and twenty-five millions of dollars. These embrace all the companies in the United States, operating on the joint stock principle. It is impossible to state definitely how many of these are operating in Venango county, but it may reasonably be supposed that much the larger portion of them are developing lands in this county. The whole product of petroleum from the States of Ohio, Virginia, and Kentucky was estimated in February, 1865, to be less than one thousand barrels per day, whilst in Venango county it cannot fall short of seven thousand barrels per day;

so that the conclusion is a rational one that the majority of these companies are connected with this county.

This capital is drawn from all sections of the country, and from all classes and conditions of life. Some of the shares are purchased in the oil valleys, and some in the most distant portions of New England. The millionaire is represented in these stocks as well as serving woman. The professional man has his certificate of stock laid by among his professional papers, and the farmer often trusts to this more than to the success of his crops. Dazzling stories are read of sudden and enormous wealth that has been secured in the oil regions, stories that are many of them true, for truth is often stranger than fiction in this region, and multitudes become intoxicated with the idea of being independent and free from care. And so the demand for shares in flattering oil companies, instead of diminishing is increasing. New companies are forming, and the business seems flattering as ever, and the mighty tide of capital is swelling, until the question is very naturally asked, "What will be the final result?"

The laws of the State of Pennsylvania, under which these companies are organized and carry on their operations, have been enacted and revised at various times, in order to adapt them as far as possible to the circumstances of the case. The act of 1864 is in the following words:—

"Three or more persons who may have associated themselves together, by articles of agreement in writing, for the purpose of carrying on any mechanical, mining, quarrying, or manufacturing business in this commonwealth, except that of distilling or manufacturing intoxicating liquors, and shall have complied with the provisions of this act, shall be and remain a corporation, under any name indicating their corporate character, assumed in their articles of association, and which is not previously in use by any other corporation or company." (Pamphlet Laws 1864, p. 1102.)

"It shall be lawful for the Governor, whenever the certificate of the organization of any such company shall have been duly executed, in conformity to the said provisions of the said acts or acts, and filed in the office of the Secretary of the Commonwealth as therein provided, to issue letters patent under the seal of the Commonwealth, declaring the subscribers to the stock of any such company, and also those who may thereafter become subscribers or holders of the said stock, to be a body politic in deed and in law, in the same manner and form as is now provided by law in other cases." (Pamphlet Laws 1855, p. 462.)

The surrender of the power of these companies may be accepted by the county courts, and provision made for the settlement of their accounts. This is done on petition of the corporation, under its seal, and with the consent of a majority of its corporators. This surrender must not, however, in any wise remove any limitation or restriction in such charter. The accounts, also, of any such dissolved company shall be settled in such court, and be approved thereby; and dividends of the effects shall be made among any corporators entitled thereto. (Pamphlet Laws, 1856, p. 293.)

Companies, by their directors, may pass such by-laws for their regulation and government as they may see fit; provided such by-laws shall not be repugnant to any of the laws of this Commonwealth, or of the United States, (Pamphlet Laws, 1854, p. 437.)

Shares of stock, for all legal purposes whatsoever, shall be deemed and treated as personal estate. (Pamphlet Laws, 1854, p. 437.)

The liability of stockholders in such corporations is set forth in these words:—

"The stockholders of any and all corporations under this act shall be personally liable for all sums of money due to laborers or operators for services rendered within six months before demand made upon the corporation, and its neglect or refusal to make payment." (Pamphlet Laws, 1864, p. 1107.)

The act of April 9, 1856, seems, however, to include materials furnished for carrying on these operations, as well as all debts contracted in carrying the product to market and selling the same. (Pamphlet Laws, p. 283.)

Companies not organized under the laws of Pennsylvania are permitted to hold lands and carry on their operations under certain restrictions. The words of the act of Assembly, of 1864, are the following:—"Any corporation, association or company not incorporated under the laws of this State, may acquire, hold and convey, not exceeding three hundred acres of land in this Commonwealth for mining purposes." (Pamphlet Laws, p. 1098.)

As many of these associations are formed under the laws of the State of New York, a brief reference to their laws may be made, although they do not differ materially from those of Pennsylvania. In forming companies there must be set forth, in the certificate filed in the office of the Secretary of State, the corporate name, objects for which the company shall be formed, amount of capital stock, term of its existence, not exceeding fifty years, number of shares into which the stock shall be divided, number of trustees and their names, who shall manage the concerns of the company for the first year, with the names of the town and county where the operations of the said company are to be carried on. (Laws of New York as amended, 1857, ch. 262.)

If these operations are to be carried on out of the State, the certificate shall so state, and also the town and county where the business is to be carried on in the State of New York, and this shall be the principal place of business within the meaning of the provisions of the act. (Laws, 1857, ch. 29, § 3.)

Trustees may call upon stockholders for all sums subscribed by them, under the penalty of forfeiting the shares of stock subscribed for, and all previous payments made thereon, if payment shall not be made within sixty days after the said demand has been properly made. (Laws, 1860, ch. 269, § 9.)

"All the stockholders of every company incorporated under this act shall be severally individually liable to the creditors of the company in which they are stockholders, to an amount equal to the amount of stock held by them respectively, for all debts and contracts made by such company, until the whole amount of capital stock fixed and limited by such company shall have been paid in," &c. (Laws, 1860, ch. 269. § 10.)

"A stockholder," however, "is not liable for debts of the corporation which were contracted before he became a stockholder." (Tracy v. Yeates, 18 Bart., 152.)

"A stockholder is liable only to the amount of his stock." (Woodruff & Beach, iron works, v. Chittenden, 4 Bosworth, 406; Garrison v. Hone, 17 New York Reports, 458.)

"Nothing but money shall be considered as payment of any part of the capital stock." (S. 14.)

The same authority is given to companies as in Pennsylvania of making by-laws for their regulation and government, with the same restriction, viz: that they shall not be repugnant to the laws of the State.

The principles on which these companies proceed are just and equitable, and if all men were just and righteous in their dealings, the plan would generally work to the advantage of all parties concerned. The business of searching for and procuring oil must necessarily be henceforth in the hands of capital. Matters have assumed such a form that the man of moderate means cannot embark in the business, for large capital is indispensable. Now, the theory in regard to these companies is that capital may be secured by many persons of small means, in connection with others of greater ability, contributing what is convenient to a common stock, and thus be prepared, in the aggregate, for competing with individuals of large capital. In theory it is like many little streams uniting their waters to form the river, on whose wide bosom commerce seeks its way to the ocean. There are many persons, too, whose tastes and avocations in life will not permit them to engage directly in this business, who, by connecting themselves with it by means of stock, may, probably, realize some profit from the connection. Persons at a distance, who may have no knowledge of the practical operations connected with the business, may, in connection with a company in which they have confidence, realize handsomely from even small investments.

As to the practical operation of these joint concerns, there has been a very great variety of experience. They need not necessarily fail—they will not always succeed. Very many of them have been prosperous beyond the most sanguine expectations of their originators and stockholders. If the site selected be a judicious one, if rich cavities be opened, and the affairs of the company be administered judiciously and faithfully, there is every prospect of success, with large dividends and liberal advance in the value of stock. Some of these companies have already declared dividends of from fifty to two hundred per cent. per annum. And some of these associations have been managed most admirably. In fact, some of the best business men and financiers in all the land have been connected with them, who have brought to bear their ripened experience, their practical knowledge, and their stern and inflexible integrity, in placing these companies on the best possible footing.

When a company has become possessed of a valuable territory, and has succeeded in developing it successfully, and managed it judiciously and honorably, its character becomes established, investment in its stock becomes safe, and its profits liberal.

But in stock companies, as in other things, there is the good and the evil—the true and the false. In all the details of business the case is the same. The spurious bank note goes on its way with the genuine, and often it is difficult to distinguish between them; yet the existence of the counterfeit does not condemn the genuine note, nor throw discredit upon the banking system. Beyond all question, there are spurious and worthless stock companies. It could not be expected that it would be otherwise, for all men are not what they should be.

There may be two classes of unreliable and worthless companies. A company may be organized in good faith, and all its operations conducted with integrity and a desire to secure the best interests of the stockholders, and yet fail of realizing the expectations of those connected with it. The lands may not prove productive. Instead of pumping and flowing wells, there may be nothing but "dry diggings;" and the result will be, that either assessments must be resorted to in order to develop new territory, or the company must disband, with the loss of all that was invested, and possibly a burden of debt to be liquidated in the best manner possible.

Again, it is a possible thing for the whole matter to be mixed up with unfaithful dealing on the part of the managers. Worthless lands may be procured, or a small extent of promising land, with a large extent of mere valueless soil appended, and the stock disposed of without any earnest honest intention of realizing for the benefit of the stockholders. The whole matter is carried forward in unfaithfulness and fraud, and the result is disaster and disappointment to the stockholders. There are numerous safeguards thrown around this business by the law, but where there is a disposition to do wrong, the opportunities are not wanting; besides, the very nature of the business precludes the large number of stockholders from knowing much about the way in which operations are carried forward. The company is, perhaps, organized in New York, or Boston, or Philadelphia. The work is carried on in Venango county, and the stockholders are, perhaps, scattered over different States in the Union, and all must ordinarily be left to the trustees and managers of the company.

Under these circumstances, then, it is not strange that many a hardly earned dollar finds its way into the rock, or what is worse, into the coffers of the sharper, to come not back again. Nor is it strange that many a brilliant dream of wealth and competence vanishes as the dreams of the night, leaving the dreamer to write "Ichabod" on his faithless certificates of stock.

But all this does not condemn these companies in their principles and practical operation. Very often stockholders in worthless companies purchase stock heedlessly and recklessly, and have but themselves to blame for their misfortunes. In this matter, as in every other connected with the practical business of life, care and caution must be exercised; dazzling promises must not be trusted; inflated advertisements must be believed sparingly, in the absence of positive knowledge either of the location, or of the integrity and judgment of the managers of the company.

Nor should extravagant expectations be indulged in, under the best circumstances. This mad and eager haste to be rich induces many to invest all, and more than they really possess, in oil stocks, under the delusive notion that sudden fortune is within their grasp. Among the tens of thousands who are mingling in the oil business in some form or other, it is manifestly impossible that all should become wealthy, for it requires a vast accumulation of money to constitute wealth in these days of petroleum.

Yet, after all, with reasonable care and judgment, and reasonable expectations in regard to a return, small investments in this promises as favorably as any other branch of business. And where there is but a small capital at hand, investment in joint stock companies, when made with care and caution, is the best way of approaching the matter.

At the present time, companies that pay dividends at all, usually declare them monthly. There is a kind of pressure in this that does not always work to the advantage of the company or the stockholders. The great majority of these companies are struggling for popularity and for life itself; and, in order to establish confidence and secure a respectable footing, it is supposed desirable to declare frequent dividends, conveying the idea that the company is in a flourishing condition. This may or may not be the case. Companies that are in a condition, as many of them are, to declare such dividends without straining their capacity or sacrificing their product when the market is dull, may safely do so; but there is a temptation in the way of feeble companies to declare monthly dividends, when, in order to do so, they are obliged to force their product on the market at a very low price, and thus work serious injury to their interests. Still, the demand is for frequent dividends, and for stock that never fails to be reported as paying monthly, even though semi-annual, or even yearly dividends would conduce more largely to the prosperity and general interests of the compan!

## CHAPTER XX.

### ORIGIN OF PETROLEUM.

AND now where shall we look for the origin of this treasure? From what elements is it elaborated? We cannot go with the great Chemist to his laboratory, and look upon the ingredients, and notice the treatment employed there. We cannot notice the number and volume of the retorts, nor look upon the mighty furnace fires that promote the distillation, nor can we tell when this mighty supply was laid up in the rocky tanks. Science, although denominated the "star eyed," cannot pentrate the mighty strata of everlasting rocks that lie beneath us, and reveal to us those mysteries of nature. There is a bound that the Almighty has placed when he says, "Hitherto shalt thou come, but no farther."

"Nec scire fas est omnia."

Nature is God's mighty domain; we may traverse a part of this domain, but not the whole extent, for we are finite, and the resources of our minds limited to a narrow horizon. God alone is great! "There is a path that no fowl knoweth, and which the vulture's eye hath not seen: the lion's whelps have not trodden it, nor the fierce lion passed by it. He putteth forth his hand upon the rock; he overturneth the mountains by the roots. He cutteth out rivers among the rocks; and his eye seeth every precious thing. He bindeth the floods from overflowing; and the thing that is hid bringeth he forth to light."

Nature has her mysteries. The earth has her great secrets. But over all a God of wisdom and goodness presides. Age after age has rolled by, change after change has agitated the history of Time, as forms of beauty have been moulded and marred, as songs of joy have been sung, and requiems of sadness chanted in the great highways and quiet by-paths of life, the living of bygone ages are slumbering quietly in the dust, and the living of the present are hurrying to the same "pale realms of shade." The nations of antiquity have passed off the stage with all their grandeur and littleness, and the nations of modern times are surging and dashing to and fro, like in the wild chaos of ocean's storms.

We go back to the history of time, ere "the morning stars sang together" their joyous refrain as earth emerged from chaos, and there is still a history—a strange, a wondrous, a mysterious existence. But this history is not chronicled in the annals of time; it is written not on pillars of brass, nor leaves of parchment, but upon the mighty leaves of the everlasting rock that make up the volume hidden beneath the earth's surface. This history is written by the same finger that graved the sublime words of the Decalogue on tables of stone upon Mount Sinai.

"The testimony of the rocks" assured us that stupendous changes were working in the earth's constitution before there were created eyes to look upon them, or rational intelligence to become conscious of them—that age after age rolled by under the guidance of the great Governor of all things, all tending to the one great sublime point when earth should emerge from her darkness and gloom, and become a fit abode for man—"Of all God's works the latest and best," a fitting theatre for the great tragedy of all the ages, and all the universe. "Mercy and truth meeting together, and righteousness and peace embracing each other!"

And during all the changes that have elasped since man was placed upon earth, a strange, mysterious work was perhaps going forward beneath us in the earth's bosom. A great dream of science, but perhaps an earnest, glowing reality suggests, that when God's almighty power was gradually preparing the earth for man's dwelling place, rolling away the curtains of darkness, forming channels for oceans and rivers, and heaping up as barriers the mountain chains of earth, his eternal prescience of man's coming need induced him to bury deep down in earth's subterranean recesses the imperfect vegetable organisms of a pre-Adamite state, that in the ages to come, coals, and oils, and gases might be drawn forth to supply his wants.

This is not mere theory. The hand-writing of nature on the rocky pages beneath, assure us that such was the case. The carboniferous age as revealed to us in the mighty coal strata of the earth assures us that so far, at least, we are drawing our supplies of fuel from the dead organisms of the mighty past; and analogous reasoning would lead us to the conclusion that our present supplies of petroleum are drawn from the same wonderful source, and proceeding from the same mysterious origin.

We find in the coal deposits traces of ferns and leaves of gigantic stature and proportions. Casts of huge boles of trees are found among our fossils, inducing the belief that in some bygone age quantities of vegetable matter, absolutely enormous, were produced on the earth's surface. Now the counsels of the Almighty are one and uniform. From the moment when this world assumed a place in his infinite plan of creation in chaotic confusion, down to the dawn of time, and thence onward to the final epoch in its history, when it will be purified by fire, and made beautiful in purity and holiness, the design was one, to prepare earth for a habitation for man, and the working out of grand and glorious results. The wonderful production of vegetable matter in a bygone age, as the material from which coal and petroleum was to be produced, would be as much a part of that plan as the drying up of the earth's surface to prepare the way for the creation of man.

It was no mere accidental circumstance that this vegetable deposit was changed to coal and oil, nor was it a merely fortuitous event that in these last years these stores of wealth were brought to light. It was the time appointed in the eternal counsels for their appearance. It was the fulfilment of the word of life, that earth should supply abundantly the wants of all the creatures moving upon its surface.

Judging from the testimony of the rocks there was a period in the eons of God, when the earth was literally burdened by immense masses of vegetation, when fronds were mighty trees, when huge foliage extended in every direction, and when reeds and rushes were so gigantic that, compared with them,

> "The tallest pine
> Hewn on Norwegian hills, to be the mast
> Of some great admiral, were but a wand."

Stillness and quiet reigned around. Every thing was favorable to the production of this rank, luxuriant growth of vegetable matter.

But revolution has been characteristic of this world's history, in the periods written on the rocks, as well as since the birth of time. The age of this wonderful vegetable deposit must pass away, to give place to new scenes in the great work of preparation for man, and it is presumable that in some of these revolutions that have agitated our planet, renovating, improving, and fitting it for a higher order of life, mighty deposits of this vegetable matter were buried up amid the rocky strata, to be revolved in new forms and products.

There is a law of attraction that runs through all the Creator's works here, that, no doubt, operated in the disposition of this vegetable deposit. The black-sand on the lake shore is attracted together, and is found collected in strata, so when the mighty revolution that buried up these primeval forests was mingling all things, this vegetable matter, by its inherent attraction, segregated itself from other substances, and settled down in immense masses in particular localities, ready for the changes that were to pass upon its substances. And it may be that since the days of Adam this vegetable deposit has, under certain circumstances, been undergoing the process of destructive distillation in the hidden regions beneath. In this process heat would not be wanting: it is furnished by the natural constitution of the earth.

Says Professor Hitchcock :—

"Whenever in Europe or America the temperature of the air, water, or rocks, in deep excavations, has been ascertained, it has been found higher than the mean temperature of the climate at the surface, and experiments have been made at hundreds of places; it is found that the heat of the earth increases rapidly as we descend below that point in the earth's crust to which the heat extends. The mean rate of increase of heat has been stated by the British Association to be one degree of Fahrenheit's Thermometer for every forty-five feet; at this rate all the known rocks in the earth would be melted at a depth of sixty miles."

But we need not think it necessary to descend thus far in the earth's crust to find the phenomena of great heat. It is often found near the surface. It bursts forth in melted rocks from the summits of mountains. Volcanic action is seen and felt everywhere. Through fissures in the primeval rocks this heat may find its way to the buried strata of vegetable deposits, and produce all the action that is necessary for the carrying out of the theory of vegetable origin.

Here then, we have all the conditions necessary to the production of petroleum. The vegetable deposit was made amid the rocks, we know not when; internal heat, with perhaps chemical action, has been decomposing that matter, and setting free its gases; these again have been condensed as they approached the surface, and have filled up the cavities, and accumulated amid the rocks, until, in these last days, the earth has literally poured out her unctuous treasures.

Again, as to the vegetable origin of petroleum, or rather that coal and oil have the same origin, we have these facts; oil almost identical with petroleum is manufactured from coal. A single ton of cannel coal placed in the retort will yield forty gallons of oil very similar to the product of the wells, and the simple agent used is heat in a confined position. Here is the proof from analysis. On the other hand we can manufacture coal from oil as it comes from the wells. In the process of distillation and refining, when the process is carried to the last degree, a substance is found in the bottom of the still strangely like anthracite coal. It lacks but the one process, that is pressure, to make anthracite coal. Here we have the proof from synthesis. And to strengthen this theory, we find coal of two different varieties in various parts of our land. One is the bituminous, that is found chiefly on the western slope of the Alleghenies, soft and full of fatty matter; the other on the eastern slope, hard, dry, and destitute of any oily matter, showing evidently that the action of heat has passed upon it, driving off its gasses, and leaving it in much the same condition as the bituminous, after the oil is extracted. Whether our present oil deposits are from the distillation of the coal beds on the eastern slope of the Alleghenies is another question. Most probably, however, they are due to a work going on beneath us, or further toward the southwest, whither the inclination and trend of the rock strata would seem to lead. A reasonable theory would seem to be that in the entire region of country where we now find the oil valleys there was at one time an internal sea, that after the great deposit of vegetable matter, in some grand revolution, a mighty deluge of sand swept over it; still further changes brought the mud deposit, then the sand again, until, at the final settling down of the different strata, a hardening process took place through the agency of heat and pressure, until we have now the diffierent strata of shale and sand rock, resulting, so far as discovered, in seven alternate strata of these two general characters.

In this hardening process we would naturally expect to find the rock twisted or bent, broken into fissures and seams, and even in cavities of great extent. This would be the case particularly if these changes were accompanied by volcanic action. The brick is of much smaller proportions after it is burned than before, as it has all its moisture driven out by heat; so the space occupied by the hardened rock must be less after it has been dried and solidified by internal heat. The product of the contraction would be crevices, cavities, and, if volcanic power be superadded, great caverns distributed here and there, that would be fit receptacles for the oil when distilled from the vegetable matter below.

Before the discovery of flowing wells, it was generally supposed that the oil was found running in slender channels and leaders through the rock, like water veins near the surface. But their spasmodic overflows have demolished this theory. The mighty fountain can only proceed from some great reservoir. The "burning well" flowed twenty thousand barrels in a comparatively limited space of time, and others have flowed steadily from one to three years. We can only account for these phenomena on the supposition of great reservoirs in the broken rock, more or less connected with each other.

But we need not necessarily proceed on the supposition that in any case these large caverns have been actually tapped. The probabilities are that they have not been reached in any instance, for the cases are rare where the bit has sunk any great distance when a vein of oil has been reached—in no case, perhaps, more than from six to twenty inches. It is probable that simple veins or crevices are tapped, that lead to the grand reservoir far beneath. In some cases vertical crevices may be struck that are of considerable extent, and in others horizontal veins, that run in various directions, until they lead to the fountain of supply. These veins are of all possible sizes—some very minute, furnishing the simple "show of oil," that is at times so deceptive to the workmen, and sometimes like the jugular vein, that leads to the centre of supply.

Such crevices and veins are found in the different strata of the rock in boring, and the probabilities are that in the rock near the source of supply they become larger as the space occupied by the vegetable deposit becomes smaller. The consequence would be, then, that the gasses and oil would rise and fill these cavities, and find their way upward until they would eventually present themselves at the earth's surface.

These cavities being at different depths, and the veins running in every possible direction, accounts for the fact that wells are found yielding oil at different depths in the same neighborhood, and even different qualities of oil; and also, for another fact, that wells within forty feet of each other do not in the least interfere. It is only when the same horizontal vein is tapped by different parties that the interference takes place.

But the different qualities of oil found at different depths shows that the product is due to distillation. At Franklin and on French creek it is found nearest the surface. At Evans' well, at the depth of seventy-two feet, the density is 29° Baume; other wells in the vicinity, but deeper, yield 32°, while the deeper wells of Oil creek are from 40° to 46°, showing that as we descend amid the rocks the oil assumes a point nearer a gaseous state, containing more naptha and a less proportion of heavier substances.

The rock where this supply is formed seems to belong to the Devonian age, but where it had or has its origin we cannot determine. The project has been suggested of sinking a shaft down to the region where it is found, of such size that the character of the strata might be determined by ocular demonstration, descending literally among the rocks, and looking at the oil in its hidden home; but the project is surrounded by insuperable difficulties, arising from the ingress of water, and the loose, open nature of many of the strata. Should the "Diamond well," referred to in another chapter, prove of practical value, it will add greatly to the knowledge we have already attained in regard to the structure and character of the rocks.

So far all is left to speculation. The hidden path yet remains unexplored. It may always remain so; but we have the great fact of Divine Providence, in the rich and copious supply, that is nevertheless valuable because it flows from an unknown source, and comes to us through unexplored channels. The practical fact is before us—we may amuse ourselves with theories as we may think best.

---

## CHAPTER XXI.

### PERMANENCE OF THE SUPPLY.

AND here another question of great and weighty importance arises in regard to the supply of petroleum. Will it be permanent? Is it laid by in sufficient quantities, either as oil, or the material from which oil is distilling, to supply the wants of man through all coming time? These are important inquiries, and cannot be fully answered; yet we can draw light from other of the resources of nature, and from the history of other oil deposits throughout the world, as well as from what we know of the structure of the rocks in our own region and the general operations of Providence.

We are often disposed to distrust Providence, when there is no ground to doubt His goodness and faithfulness. There is as little room to doubt the continuance of this rich supply as that of other natural supplies. There was a time when rational, intelligent men feared that the coal would all be exhausted, and the world be left to perish for want of fuel. This was so much the feeling in England a few years ago that a commission was appointed to make a geological survey, and report on the subject. The result of the survey and investigation was, that there was coal in the mines of England alone, to answer all the purposes of the nation for many thousand years, or until England should no longer need fuel. So it was in our own land before geological surveys had opened up the resources of our country. Its existence was not suspected in many places where it now abounds, and its extent is greatly magnified to our observation in regions where we before knew of its presence. The wants of the world, and the pressure of these wants, lead to investigation, and this results in discovery.

The light of the past is surely worth something in this matter. We read of it first on the plains of Shinar four thousand years ago; it is found in the same locality at this day. We read of slime-pits, or petroleum springs, in the days of Abraham—they have not yet disappeared. We read of the building of Babylon in the sacred book, and Herodotus, Diodorus the Sicilian, Josephus, and other profane historians tell us that its walls were cemented with bitumen, and that this bitumen was brought from the banks of the Issus, a tributary of the Euphrates. The testimony of modern travelers is, that this substance is found in large quantities on the Issus at the present time. No doubt there were wise Babylonians in the days of Nimrod, and Semiramis, and Nebuchanezzar, who predicted that the pitch on the banks of the Issus would soon be exhausted, as they were drawing so lavishly upon it to build the walls and towers and monuments of the great city. Still they went on building and enlarging and beautifying that wondrous city, and drew all the petroleum that was needed, and still the supply did not fail; and it is bubbling up there freely now, and, no doubt, some new Nimrod might go on and rebuild the fallen city, and another Semiramis enlarge it from the same material, and still not exhaust the supply.

Petroleum, in much the same form in which we find it in our own valleys, was gathered in Asia, on the banks of the Irrawaddy, as long back as we have authentic history to guide us. The Burmah wells have been famous for more than two hundred years. They have been yielding about one thousand barrels annually, under the unfavorable circumstances connected with them, and yet manifest no signs of failure.

If we come down to more modern times, we find petroleum, in the form and under the name of Barbadoes tar, produced largely in the West India Islands. It has, for many years, been largely exported to almost all foreign countries for medical purposes, without, as yet, making any impression on its quantity, or reducing it in any perceptible degree. Humboldt, writing in 1799, speaks of the product of oil in the same island as being immense, in the form of naptha, (most probably the form in which we find the oil here) spreading itself over "a large surface of the sea." Petroleum is found in the West India Islands in large quantities still; a flow of sixty-five years, that was, no doubt, constant in its quantity, has not exhausted it.

The same might be said in regard to the yield in our own valleys. We can trace its history here for one hundred and twenty years. It was as plenty then on the surface as now. It could be gathered during all these years in quantities as great as the limited machinery employed would admit. Vast quantities made its escape every year that was not noted. It bubbled up through a thousand cavities in the beds of our streams, was beautiful for a moment as it spread itself upon the water, and then vanished forever. It oozed forth from small crevices in the rock, and found its way to the surface of the ground, and evaporated, leaving little signs of its presence. The gases dispersed themselves in an invisible form through the atmosphere, and were unnoticed. Still there was no inconsiderable amount of petroleum produced from year to year. Yet we find the store-houses to be amply filled.

It may be said that never before was there such a drainage upon the fountain as now, and that in the history of the ancient wells alluded to, modern American genius and enterprise had never been brought to bear. Even so, but the same argument might be brought to bear in the coal supplies, and made to show that they would soon be exhausted, since ocean steamships and modern machinery of various kinds are making such havoc with their stores. The only difference consists in this, as far as the force of the argument is concerned: The coal beds lie near the surface, and a very clear approximation can be made as to their volume, and the extent of their resources; whereas petroleum lies far down in concealed caverns and rocky recesses, where no human eye can penetrate, and no human hand can gauge its quantity or compute its volume. Yet there is no more reason to suppose that the latter will in time be exhausted than the former.

The whole history of the past may be searched in vain to find evidence that would lead to a fear that the oil deposits will eventually be exhausted. Not a single article that has been supplied for the comfort or welfare of man has ever been withdrawn by the hand that gave it. New articles are brought to light again and again, new uses for articles already well known are frequently discovered, but not one instance where an article has been largely developed, and of great practical utility, is recorded as having been withdrawn from use by a failure in the supply. It would be preposterous, then, to suppose that an article of such general utility, and capable of being brought to bear in so many of the operations of life, should be thus exhausted, and cease to be known among the great staples of the world.

As an example of the inexhaustible nature of all the really useful productions of the world, mention might be made of tin. For a long time it was found only in Cornwall, England. The veins that contained it were comparatively small and insignificant. It was secreted in hard granite. Mining operations were carried on with great difficulty and expense. The territory where these small veins were found was limited, and it seemed that the supply must soon be exhausted. Yet those Cornwall mines have been supplying the world with tin for more than twenty-five hundred years, and are not yet exhausted. There is evidence that this metal, more highly valued then than now, found its way to the Temple of Solomon. It finds its way still into every temple and household throughout the civilized world.

So it will be with petroleum. If the demand is greater than for tin, the sources of supply are larger, the evidences of its dissemination wider and more extensive. There is really no more reason to apprehend a total failure of this article than of the wood of the forest, or the coal of the mines. We may feel sure that it has taken its place among the valuable and indispensable articles that belong to the human family for all coming time.

We have already noticed the supplies that were laid down in the unknown ages of the history of the world, and, from the character of the vegetation of that age, we may well imagine that that supply would be copious. We do not find in the oil-bearing rock any fossil remains, because the region cannot be exposed to ocular view. The rock that is brought to the surface in the sand pump is beaten fine almost as the dust of the balance. Its substance can be detected, but not its peculiar characteristics. But, on the surface of the rock, and in the strata that crop out from the hills all over the oil region, we find remains of vegetable deposit. Its paleontology is peculiarly rich and abundant. We cannot doubt, then, but that the deposit of vegetable matter beneath is plentiful if not inexhaustible.

We have indubitable evidence that the structure of the rock is broken and open. The number of water veins is evidence of this. The mud vein spoken of in a previous chapter affords corroborative proof, inasmuch as it suggests that not only water but earthy deposit has percolated through the rock to a great depth beneath the surface. And the fact that, with scarcely an exception, the operation of boring has revealed seams and crevices, confirms the opinion that both the sand-rock and shale are broken and seamed, and filled with cavities. Now, these cavities may be of very great extent, deeper than the drill has ever pierced; these cavities may be larger and more extended than they appear to be in regions partially explored. No doubt they ramify in all directions, reservoir connected with reservoir, and these with connections that extend upwards through a stratum not so largely seamed and broken, until they approach the sphere to which boring operations have already extended. This latter region appears to be full of comparatively small cavities, and to be reticulated by leaders that run in almost all directions, but growing smaller and more confined as they approach the surface of the rock.

If this theory be correct then, and the nature of the rock be more open and broken and upheaved as we descend, we may naturally expect that the cavities far beneath will be capable of furnishing an inexhaustible supply.

But we have no evidence that the formation of oil has ceased in the regions below. We know that the mighty furnace fires are still kept up, and there is reason to believe that the store of vegetable deposit was at the first ample, and why may we not suppose that the mighty retorts in the rocks below are still kept in active operation? There is no limit, surely, to Omnipotence. And if we take the mode of his operation, as we see it carried forward on the earth's surface, as a criterion by which to judge, we shall be strengthened in the belief that the same course of production and supply is carried out in the regions below. As a general thing, every year produces its own harvests. The ground is constantly changing. Through the action of air and rain and frost and the running streams, the soil is deepened and made productive. The forests grow and decay, assisting in the same design. The coral insect, by its minute yet ceaseless labors, raises islands and peninsulas from the ocean's depths, to add to the aggregate of tillable land, as the inhabitants of earth increase. The currents and streams of the ocean preserve its waters, while the winds and storms purify the air and render it healthful. The waters evaporate and pass upward, to return again in the form of rain and snow, to refresh the earth and minister to the comfort of its inhabitants. There is a ceaseless, constant change going on all around us, in order to work out grand, blessed results for the good of earth's inhabitants.

Changes analagous to these are most probably going on in the great deeps that cannot be fathomed by mortal lines, and that cannot be searched by mortal eyes. The same power operates there that is more clearly visible at the earth's surface. The same hand makes provision beneath as above, and the same activity in operation and fertility of supply would be manifest there, could we see it, as is so obvious in the times and seasons of our daily experience.

"Deep in unfathomable mines
  Of never failing skill,
He treasures up His bright designs,
  And works His sovereign will."

We need not be led by this reasoning to suppose that the supply either of vegetable deposit, or the process of distillation or production of oil, is to be found in every part of the earth's interior. At the time when the former was laid down in its place of repose, there were peculiar causes, no doubt, in operation, by which this mass of vegetable organisms was aggregated in particular localities. Mighty revolutions, no doubt, accompanied the changes in the earth's features, as it was gradually approaching its present form. In the Devonian age, when this deposit was made, as the earth's crust was heaving and agitating, rapid currents of water would naturally bear it in masses to deep basins and cavities, assisted in part by the attractive powers of its own particles. This, in turn, would be buried by the deluge of sand, and this again by mud and dark-colored sediment, that now constitute the sand-rock and shale.

In some other revolution the superincumbent mass

would be broken, and contorted, and upheaved until the vast caverns were provided as the retorts and tanks of the great Disposer; the fires commenced their operation, and the oil and gas began their upward journey. Now these fires have not yet been extinguished. They will continue until the period when "the elements will melt with fervent heat," and when "the earth also and the works that are therein shall be burned up." We have no reason to think that the supplies for its production have been exhausted, and may then reasonably infer that the process of manufacture is still going on, and that it cannot be exhausted.

But in the face of all these theories the question is asked:—"Do not wells become exhausted?" "Are there not instances on record where wells have ceased their supplies altogether, and is not this an indication that all will do so?" Wells do cease to flow, and those that have been worked by means of the pump have also become quiet, and ceased their supply. But this is no indication that all will do so. We find many trees in the forest that are dead, but that is no indication that the forests are decaying, and will soon die out. We find coal veins sometimes suddenly running out. The miner will find a "fault," as he terms it, when instead of coal he will find nothing but a bank of earth, or perhaps solid rock. But this does not discourage him. He mines through the earth or rock, and finds a new vein of coal as rich as the first. And it may reasonably be supposed that there are "faults" in oil veins. One vein may be exhausted, or one cavity may be drained that may require long years to fill up, as the veins of supply may be exceedingly small. But there may be others in the same neighborhood as rich as the first, that require only to be pierced to obtain a new supply.

But wells may cease their supply from other causes than simple exhaustion. The supply may proceed from a long, thin crevice, and this crevice may become closed by mechanical means. Sometimes there are presented thin scales of paraffine, indicating the presence of this substance in the well; this paraffine may gradually adhere to the opening in the crevice until it becomes hermetically sealed up. Or the same result may be accomplished by the settling down of sand and earthy particles from the side of the well. Sometimes when a well ceases its supply it can be resuscitated by boring deeper, but not always. The very operation of boring sometimes fills up the crevices more firmly than ever by pounding and forcing particles of sand and mud into them beyond the hope of relief.

The mere fact, then, that individual wells occasionally cease their supply is not evidence that the general supply is either failing now, or is in any danger of ultimate failure. The same failures occur at times in wells of water for domestic use, in coal mines, and in beds of iron ore. It is more particularly true in regard to the mines of silver and gold, and cinnabar in our new States. Yet these occasional failures are not regarded as evidence of any decline in the business. When a water well gives out it is dug deeper, or a new well provided. When a vein of coal or iron ceases to yield, through a "fault," the work is continued until a new vein is opened, and in regard to the precious metals, when a leader gives out, it is sought in a new place. And this doubtless will be the case in the oil business. Wells will fail, sometimes to be resuscitated, at others to be abandoned. New wells will be opened in the same neighborhood with abandoned wells, better than the former ones, and so the supply will be kept up from time to time, with greater or less constancy.

The region of oil producing land will no doubt be much extended. Profitable development on the Allegheny, Oil and French creeks will stimulate the matter of prospecting in other parts of this and other counties and States, resulting in the discovery of other deposits, that will assist in keeping up the supply, and meeting the increasing demand that is springing up in all portions of the world.

It is not at all unlikely that the annual supply may vary from year to year. This has been the case already. During particular years when vast numbers of flowing wells were opened the supply was immensely increased; and when these decreased, and resort was had to pumping, the aggregate supply was materially reduced. This may be the case in the future. Peculiar causes may induce a large amount of exploration that may result in large additions to the stock of oil produced, and, on the other hand, different causes may lead to a different state of things; attention directed to other branches of business, and for a time neglect of this; but the general result, no doubt, will be that the supply of petroleum will be kept up while there is need of artificial light to carry on the operations of life, whilst any of the purposes to which it is now applied remain to be fulfilled, and until the present organization of society and the world shall have been finally changed. To go upon any other supposition would be to suppose that the course of nature, and the operations of Providence would be changed, and God's wisdom and power cease to be adequate to the supply of the wants of his creatures.

# CHAPTER XXII.

## GROWTH OF THE OIL BUSINESS.

THE history of the growth and expansion of this trade is the most remarkable in all the operations of men. At first, as we have had occasion to notice, it was small and almost contemptible. Considered simply as a drug, valuable, but of little importance in trade, very little attention was paid to it. It was considered a curious product, a lusus naturæ almost, but no one imagined that it would ever assume proportions so gigantic as those that characterize it at the present time. First a few bottles would find their way out of the immediate neighborhood where it was produced. Then a quantity somewhat larger, but sold in the shops under a fictitious and foreign name, as though the adage "a prophet is not without honor, save in his own country" was applicable to the productions of the earth. In the course of half a century after the settlement of the oil valleys by civilized people, it had not gained a particle either in the way of appreciation or pecuniary value. The difficulty of forcing itself upon the market seemed as great as that of forcing its way up through the rocks. But both ends seem very readily attained at the present time.

Previous to the year 1859 hardly a barrel of petroleum found its way to market, but during that year the great highways of trade were opened to receive it. The manufacture of oil from coal had been the means of preparing the way for its advent. The matter of refining and deodorizing coal oil was brought to bear on the new article so suddenly thrust upon public attention, and at once the demand was constant and permanent. So it has continued to be until the present time, with the exception of a short period when the flowing wells seemed likely to deluge the land with oil at a merely nominal price.

Perhaps, at the present time, there is not more oil produced than two years ago. But withal the business is growing. The product then was in excess from causes already stated; now it is steady and regular, not in excess of the demand, and in every respect healthy and prosperous.

It now occupies a very large place in home consumption. From the various uses to which it is applied, and which are multiplying from year to year, the demand has been steadily increasing. The demand comes from every State and Territory of the Union, except those whose markets have been closed during the stern arbitrament of war; and now this is ended, there will soon be a new source of demand, and a new field to supply. At first it was regarded simply as an illuminator, now it is wanted for very many other purposes, and next year perhaps it will be applied to many purposes not even imagined at the present time. And just in proportion to the number and magnitude of the uses to which it is applied will the demand increase provided the sources of supply are kept up.

From the best information that is at hand the consumption in our own land during the year 1864, must have been one million two hundred thousand barrels. The principal portion of this was refined, and used as an illuminator, the remaining portions in the various arts and practical operations of life.

The growth of the export trade has been steady and regular. At first it was difficult to introduce petroleum to the notice of Englishmen. A cask of the article at Liverpool was clothed with almost as many terrors as a supposed infernal machine. But the English are a practical people, and understand their own interests as well as any other nation, consequently they soon began to value the American production in something like its true light. Next to England, France appears to appreciate the new commodity, as certified by her imports, reaching during the year 1864, the amount of four millions six hundred thousand gallons.

The following table will be interesting and instructive, as exhibiting not only the growth of the trade, but the universality of the use of petroleum throughout the civilized world. The exports from New York alone are given to their different points of destination; from the other points the destination is not given; probably no new points are embraced in the shipments from these ports.

| FROM NEW YORK | 1863. GALLONS. | 1864. GALLONS. |
|---|---|---|
| To Liverpool | 2,156,851 | 734,755 |
| London | 2,576,331 | 1,430,710 |
| Glasgow, &c. | 414,943 | 368,402 |
| Bristol | 71,912 | 29,134 |
| Falmouth, E | 623,176 | 316,402 |
| Grangemouth, E. | 425,334 | ——— |
| Cork, &c. | 1,532,257 | 8,310,362 |
| Bowling, E. | ——— | 87,164 |

| From New York | 1863. Gallons. | 1864. Gallons. |
|---|---|---|
| To Havre | 1,774,890 | 2,324,017 |
| Marseilles | 167,896 | 1,982,075 |
| Cette | ——— | 4,800 |
| Dunkirk | ——— | 232,803 |
| Dieppe | 46,000 | 79,581 |
| Rouen | 143,646 | ——— |
| Antwerp | 2,692,974 | 4,149,821 |
| Bremen | 903,004 | 971,905 |
| Amsterdam | 436 | 77,041 |
| Hamburgh | 1,466,155 | 1,186,080 |
| Rotterdam | 757,249 | 522,926 |
| Gottenburgh | ——— | 33,813 |
| Cronstadt | 88,060 | 400,376 |
| Cadiz and Malaga | 33,284 | 55,674 |
| Terragona and Alicanti | 33,000 | 16,823 |
| Barcelona | ——— | 25,500 |
| Gibralter | 308,450 | 69,181 |
| Oporto | 2,339 | 17,474 |
| Palermo | 57,115 | 7,983 |
| Genoa and Leghorn | 399,674 | 635,121 |
| Trieste | 3,000 | 165,174 |
| Alexandria, Egypt | ——— | 4,000 |
| Lisbon | 64,662 | 167,195 |
| Canary Islands | 5,125 | 3,358 |
| Madeira | 400 | ——— |
| Bilboa | ——— | 2,500 |
| China and East Indies | 36,942 | 34,388 |
| Africa | 12,230 | 25,195 |
| Australia | 304,166 | 377,884 |
| Otago, New Zealand | 5,500 | 10,810 |
| Sydney, New South Wales | 43,012 | 97,880 |
| Brazil | 160,152 | 140,677 |
| Mexico | 69,451 | 112,986 |
| Cuba | 356,436 | 418,934 |
| Argentine Republic | 24,470 | 28,260 |
| Cisalpine Republic | 117,626 | 78,552 |

| FROM NEW YORK | 1863. GALLONS. | 1864. GALLONS. |
|---|---|---|
| To Chili | 66,550 | 92,550 |
| Peru | 256,467 | 169,061 |
| British Honduras | 440 | 6,072 |
| " Guiana | 15,104 | 7,881 |
| " West Indies | 60,931 | 70,976 |
| " North American Colonies | 16,995 | 28,902 |
| Danish West Indies | 31,503 | 8,463 |
| Dutch " " | 12,143 | 26,638 |
| French " " | 9,104 | 16,020 |
| Hayti | 12,064 | 7,088 |
| Central America | 456 | 993 |
| Venezuela | 15,405 | 29,583 |
| New Grenada | 107,837 | 57,490 |
| Porto Rico | 59,439 | 20,626 |
| Total gallons | 18,542,586 | 26,281,059 |

The following was shipped from other ports during the same years:—

| | 1863. GALLONS. | 1864. GALLONS. |
|---|---|---|
| From Boston | 2,049,431 | 1,696,308 |
| Philadelphia | 5,595,738 | 7,760,148 |
| Baltimore | 715,896 | 929,671 |
| Portland | 342,082 | 70,672 |
| Total | 8,703,147 | 10,456,799 |
| Add New York export | 18,542,586 | 26,281,059 |
| Total export from United States | 27,445,733 | 36,737,858 |

Same in 1862 .......... 10,887,701 gallons.

From this table we see the dispersion of the trade. Europe, Asia, Africa, and South America, as well as Islands in the Atlantic, Pacific, and Indian Oceans, come in for a share of the exports, and, as it has just been introduced into many of these places, the hope is not an unreasonable one, that, as its value becomes known, there will be a largely increased demand from year to year as long as the supply continues. In most of the places enumerated in the table petroleum was unknown four years ago, except perhaps as a medical agent, or as a natural curiosity. In 1861 forty casks were sent to France as a curiosity; the next year some four thousand casks were sent there by way of experiment, now this country competes with England in her demand. The foreign demand for illuminating purposes alone would be very great in a short time, but as new discoveries are made, and new uses developed for the product a very rapid increase in the demand may readily be expected, that can only be qualified by two considerations, to wit, large discoveries in foreign countries, or a large falling off in our own country.

As to the former, the idea is not an extravagant one that the same means being employed, and the same energy put forth that are brought to bear in the development of the trade here, there may be like success. Still, we are not sufficiently acquainted with the geological features of foreign countries, where this product manifests itself, to arrive at any opinion in the matter. As to the other modifying condition, the falling off in the supply from the oil region in this country, there is no ground to indulge in fear; on the contrary, every indication points to a large extension in the sources of supply.

During the past year new capital has been seeking investment in the oil region to an extent absolutely astounding. The very circumstances of the country are favorable to the growth of the trade. Capital is abundant. Hopes of increase are large, and the petroleum land seems the land of promise. Lands that were valued at thousands a few years ago are now valued at millions. Capital that was formerly seeking investment in railroad stocks, gold and silver stocks, iron and coal stocks, is now rushing tumultuously into the oil valleys, finding employment there, with a fair show of success. It is entirely safe to put down the capital connected with the petroleum business in the United States at four hundred and fifty millions of dollars. Of this more than one half is connected with Venango county.

Now, as a very large amount of capital has been invested in oil lands, or lands supposed to overlie oil, and this supplemented by other capital set apart for the development of these lands, the supposition is a reasonable one, that a large amount of valuable oil lands will be rendered productive in the course of the current year. The same energy that has characterized the past, and that has been attended with so much success, will be brought to bear in the development of these new lands, and will, no doubt, be attended with similar success.

Nor is the idea a correct one, that these new lands are probably worthless. No doubt many portions of them are. But, as explorations proceed in new portions of Venango county, on the tributaries of the streams where operations have hitherto been carried forward, they have been attended by the most gratifying success. In many places the new territory has rivalled if not excelled the old. And just in proportion as the field is widening, and capital increasing, will the trade increase, until, at no distant day, it will surpass in magnitude and importance any other branch of trade in the whole land. It is already pressing hard upon the iron and coal interests. It will soon overleap in importance the gold and silver products of the western States and territories, and the entire yield of cotton in the southern States in their most palmy days.

The business is in a more flourishing condtion now than it has ever been before; new capital is entering into its development, and parties engaged in it are quite satisfied with its progress. There are, of course, drawbacks and discouragements; these belong necessarily to all enterprises; but the general features of the entire business are most encouraging and promising; and, unless all rational indications usually manifested in the progress and development of trade should fail, the petroleum business has before it a grand and brilliant future.

---

## CHAPTER XXIII.

### PRESENT ASPECT AND IMPORTANCE.

THE petroleum trade is singular and unparalleled in the world's history. Like Minerva, in Grecian fable, springing mature and full-armed from the brain of Jupiter, it has sprung in almost mature strength and vigor from the earth's bosom. It has worn no swaddling bands. It has required no gentle nurture, no fostering care, to enable it to eke out an existence until time should give it strength and harden its muscle, but has leaped at once into the arena of the world's traffic, and is now the peer, if not the superior, of all other branches of trade. In less than one year from the moment of its inception, it has fairly eclipsed the whale fishery, gray with time, and strong through the energy and vigor with which it has been prosecuted. And who can measure its extent in the future, since it can only be limited by the sources of the supply flowing in the depths of the laboratories of the Great Chemist? No doubt the stores of earth and the supplies of time are all limited, but it is only by the wants of His creatures, and not by the power of the Almighty. If earth, on its surface, "bringeth forth herbs meet for them by whom it is dressed," we can have as little doubt that deep in its bosom there is an exhaustless supply of minerals, to neutralize its cold, to dissipate its darkness, and to contribute in various ways to the comfort and enjoyment of its inhabitants. And among these gifts and resources petroleum stands forth in the front rank. It is God's greatest gift of the nineteenth century.

The trade is flourishing as no trade ever did before. A few years ago petroleum was not found in the price lists of the market. A few years after it was first begun to be developed it was with fear and trembling that capitalists embarked in its operations; now it is quoted as regularly as gold. It is found on the stock board—nay, it has a stock exchange of its own, a thing that cannot be said of any one other article of trade. It is attracting more attention and interest than all other branches of trade or commerce. Men of science, and lovers of nature, and explorers of her hidden resources and mysterious laws, are giving it their attention, and contributing the results of their knowledge toward its development. Capitalists are investing their funds with unsparing measure in prosecuting its details, and the very nation, torn and bleeding and suffering, looks to it as the great resource from which it is to draw its recuperative energy now that the war has drawn to a close. As yet there is not a circumstance to indicate that the resources of the trade are in the slightest danger of lessening, much less of drying up. Its appearance is that of a vigorous young tree, luxuriously covered with leaves and blossoms, and full of promise for the future.

The influence of petroleum on general trade can hardly be overestimated. With all due allowance for inflation and over-trading, and undue extension of business, its effect has been most salutary and beneficent upon the general interests of trade throughout the country. In these peculiar times it has supplied a field for the employment of capital that has had a kindly influence in all our business affairs.

Its importance to the manufacturing interests of the country is very great. New branches of mechanical business have been called into existence, and all branches stimulated to new life and energy. In the matter of "driving pipe" alone, an idea may be formed of the demand, when it is remembered that in the prosecution of a single well from one to three tons of metal pipe are driven in reaching the rock preparatory to boring; and although there are regions along the Allegheny where but little pipe is necessary, the rock lying near the surface, yet in other places it is so deep that one hundred feet are sometimes driven, at an expense of some seven dollars per foot.

Boring implements, too, have called for a vast amount of labor and material. A complete set of these is furnished at about three hundred and twenty-five dollars, and when it is remembered that the wells are numbered by thousands, each company having, as a general thing, a set for each well that is bored, the capital invested in these alone is not by any means small.

Steam engines are as numerous in the oil regions as cooking stoves in the kitchens. They are of various kinds, and as many forms as Proteus had shapes. There are several thousand now in active operation in Venango county, and hundreds are arriving every week. These engines are manufactured in various parts of the country—in Pennsylvania, New York, New England and Ohio, and in their construction a vast amount of material is required, and a great number of workmen employed.

The manufacture of "chamber," or "tubing," is an important item in the manufacturing interests of the country. Each well requires from four to six hundred feet of chamber, at a cost of from one to two and a half dollars per foot. Cable and rope for boring and sand-pump attachment must also be taken into the account, with other items that swell the amount of mechanical business pertaining to the mere matter of putting an oil well into operation, to a very large amount.

But there are other matters outside of the mere matter of boring and pumping that must be taken into the account. Some of these matters are new, others are the expansion of branches of business that are already in successful operation. In the manufacture of lamps and glassware connected with lamps, there is a very large amount of capital invested, and a great number of laborers employed. In the city of Pittsburgh there are five establishments engaged exclusively in the manufacture of lamp chimneys, turning out four thousand dozen weekly. There are also eight establishments engaged chiefly in the manufacture of lamps, and two others give their entire attention to this branch of the business.

These factories employ from one to two thousand workmen, and embrace a capital of about half a million of dollars. This business has sprung up too within the last three or four years,

The business of refining and preparing lubricating oil is one of immense extent. The former requires extensive and costly machinery and fixtures, with large quantities of expensive chemicals, the production of which gives employment to workmen in all portions of the country. Venango county alone has within its limit from ninety to one hundred refineries, with an aggregate capital of one million dollars. These refineries are springing up in other portions of this, as well as in various portions of the neighboring States.

The manufacture of barrels is a large item connected with the business affairs of the country. At the inception of the trade these were generally manufactured by hand, and in the old orthodox manner of country barrels, laboriously shaved from prime oak timber, and hooped with hickory hoops. But the demand was too great, and prices too remunerative for this business to remain in the old paths. Native energy was put forth, and new plans adopted. Good timber was becoming scarce in many parts of the country. Resort was had to the use of machinery, bringing into use a poorer quality of timber, and producing barrels with great facility and dispatch. In some factories the staves are sawed out to the proper shape and put together after being kiln-dried, hooped with iron, tested with the air-pump, and sent to the oil wells in large quantities. This brings the timber lands into higher estimation than they were held before, furnishes labor to farmers during the winter season in cutting and hauling timber, to workmen in the barrel fac-

factories, and makes a demand for labor generally throughout the oil country. But this matter is not confined to the oil region. Barrels are brought in large numbers from a distance. They are floated down the Allegheny river from the State of New York, and brought up the river from Pittsburgh below. They are brought from Ohio by teams, and from every direction producers are certain of a market at remunerative prices. For refined oil a rather better class of barrels is required, as the wastage is of more importance. These barrels are usually coated on the inner surface with a preparation of glue, or some other substance that will resist the action of the oil.

Railroads have been constructed for the accommodation of this business alone. Along the valleys of Oil creek, French creek, and portions of the Allegheny, where the citizens never dreamed of seeing a locomotive or hearing its wild scream, these have become familiar sights and sounds. On all the roads that have been constructed in the oil region, the trains are crowded both with freight and passengers. Rolling stock cannot be procured in sufficient quantity, the crossings are clogged up, and the freight depots crowded to their utmost capacity. And from present indications and future prospects, the freighting business may be enlarged almost indefinitely. Measures are already on foot for enlarging greatly these railroad accommodations in various directions.

Employment is furnished to vast numbers of teams in the oil country. From the region of Oil creek and the Allegheny alone, perhaps over one thousand horses are at all times busily employed. These teams draw the oil from the wells to the refiners, to the railroad depots, and to the steamboat landings. The usual order is to load with empty barrels, or "emptys," as the oil men call them, at the depot, proceed to the wells, and return with a load of oil. In many cases there is an economical improvement on this plan. In leaving Franklin the teamsters purchase a load of coal from the coal men, who are thronging the streets with their wagons, place their "emptys" upon the top of this, and proceed to the oil wells. Here they dispose of their coal at a handsome profit, and re-load with oil.

This is a matter of considerable importance to the farmers in Venango county. They are near the scene of labor, can often furnish their own subsistence for their teams, and make a clear profit of the greater portion of the receipts; and, in addition, can choose their own time for active effort, and be on their farms when their presence is required there. Still the labor is not a pleasant one. Clothes saturated with oil, wagons broken and dismembered, harness and horses wearing out rapidly, and a general condition of dilapidation and decay belongs to the business. If the teamster receives good wages, his experience is not different from that of others, that there is another side to the account that must be considered in balancing the ledger.

Still the discomforts in this business are not greater than in many other fields of labor, and it is perhaps more remunerative. And this labor is of various kinds. A great number of men are employed at the oil wells, in the active operations of boring and pumping. Laborers are required at the depots, in the cities, and around the wharves. A mighty army is engaged in all the processes of manufacturing the oil, and in preparing machinery for the various operations connected with it.

Its influence is felt in every State and Territory of the Union, and in all the walks of life, in the matter of labor; for, although it seems at first view, as we look at a flowing well that but little labor is necessary, yet as we follow its history from the time preparation is made for boring until the oil is consumed, we will see that a vast amount of labor is connected with it.

Its influence in the Ocean trade has been felt perhaps in two ways. First, in lessening the amount and importance of the whale fishery; second, in building up and enlarging a trade in carrying petroleum to foreign ports. The whale fishery has certainly declined rapidly during the last four or five years; yet it is probable this is due in part at least to causes outside of the oil business. From constant pursuit the whales had been driven to high Northern latitudes, as well as decreased in numbers, so that of late years the same success had not attended the efforts of the whaling fleet, that had characterized it in former years. "The diminution of the whaling fleet," says the New Bedford Standard, "has continued the past year, but the decrease is small compared with previous years, and we hope has now reached its minimum. The decrease for the year has been twenty-eight vessels, with a tonnage of eight thousand eight hundred and seventy-two. In 1846 the tonnage engaged in this fishery was two hundred and thirty-two thousand two hundred and eighteen; now it is only fifty-nine thousand nine hundred and two. In 1843 sperm oil sold for sixty-three cents per gallon, in 1864 for one dollar and ninety-two cents. In 1831 whale oil sold for thirty and a half cents; now it sells for one dollar and twenty-eight cents. In 1841 whale-bone sold for nineteen cents per pound; now it sells for one dollar and eighty-two cents."

There can be no doubt, however, that apart from the scarcity of fish, the influence of the petroleum discoveries has affected the whale oil trade, by bringing into competition with it a fluid that answers a better purpose in many respects, and is afforded at a much lower price. In fact, many of the old whalemen have thrown aside the tarpaulin and harpoon, and have entered upon the business of selecting sites and locating wells with as much avidity as ever they pursued the monsters of the deep; and have expressed themselves as well pleased with the exchange both as regards comfort in pursuit, and profit in the final result. And although the amount of tonnage employed in the petroleum trade may never equal that formerly engaged in the whale fishery, yet that which is employed will undoubtedly accomplish more; as a whale ship was often from twenty months to three years in bringing a cargo into port, while a petroleum ship would make a voyage to Europe and return in a few weeks.

The influence of this trade is felt at home. Not only are the comforts of its production felt in the immediate region of the oil valleys, but in every nook and corner of our wide land. The lamp filled with petroleum is found burning in every cottage and log cabin from the Atlantic coast to the Pacific. By its grateful light the humble artizan and the poor sewing-woman carry on their toilsome avocations, grateful for the illuminating powers, and the comparative cheapness of this providential gift. And in every part of the land, energy has been quickened, industry stimulated, and genius and talent called out in the various processes that pertain to developing wells, producing oil, and extracting from it the various substances that add to the comforts and enjoyment of society. Even in the dreary season of winter the frost and snow have interposed no barrier to its production. Eager crowds throng the public conveyances that move to and from the oil valleys. The streets and hotels in the oil region are full to overflowing, and the bustle and stir surpass that of California in its most palmy days. Earnest, eager looking men are seen everywhere with the peculiar appearance that characterizes the "oil man;" short coat of peculiar cut, waist extravagantly long, and skirt infinitessimally short; smallest possible hat; tall boots, with little appearance of pantaloons; bundle of papers in breast pocket, and sectional map in hand supposed to represent "valuable oil territory," from which a fortune is to be extracted either by pumping it from the earth, or by selling it to some other aspirant for the favors of fickle fortune. But fortune is not supposed to be fickle in these regions. She is supposed to have taken the bandage from her eyes, and to be gracious to all comers, rewarding all who approach her in the oil garb with untold wealth; regardless of the experience of multitudes who walked these valleys three or four years ago, and have left their derricks behind them as monuments of their disappointed hopes.

Its importance is seen in the altered condition of very many in the oil region and beyond who have come within the sphere of its influence. Many have become suddenly and immensely wealthy, who were formerly in very straitened circumstances. Plain farmers have become millionaires; men whose incomes were half a dollar per day a few years ago now have an income of five thousand per day. Still in very few instances has this sudden wealth appeared to have an injurious influence upon its possessor. In a very few cases it has dazzled, and blinded, and blighted all the better feelings of the heart, but generally it has not, apparently at least, led to dissipation, vice, and ruin. On the contrary, in very many noble instances it has had a happy effect in inducing to benevolent works, to acts of kindness, and to labors of love and sympathy.

In other lands the influence of this trade has been very great. It is bidding fair to revolutionize the habits, and change the associations of the people as it has already done here. The old tallow candle, and the well nigh fossilized oil lamp, will soon disappear as the petroleum lamp is introduced, and throws them into the shadow. The statistics connected with the exportation of petroleum indicate, in some degree, the estimate that is placed on this commodity in almost every country in Europe, Asia, and South America. And there can be no question but that in proportion as its value becomes known abroad the demand for it will increase in a regular ratio. It may, however, be discovered in many countries where it is now imported, still the geological features of many of the principal points of export would naturally preclude the idea of its discovery or production.

The influence of the export trade is felt in a marked degree in regulating the balance of trade with foreign countries. Our imports, either from necessity or choice, are very large, especially from England and France. During the past year, 1864, there were exported to England upwards of eleven and a quarter millions of gallons and to France over four and a half millions of gallons, affording this country advantage to this amount over our exports of other commodities during previous years, or before this trade commenced. This influence has been most beneficial to our trade during the disturbance in our pecuniary affairs for the last few years, and we may reasonably expect that for years to come we may depend upon petroleum to supplement our other exports and thus prevent the drainage of the precious metals from our coffers.

A wise and benignant Providence is hereby providing for our national affliction. When any part of the body is wounded or receives injury, nature appears to turn all her resources to the healing of the wound, and the recovery from the injury. New resources even, are opened up exactly adapted to meet the emergency and to obviate the difficulty. The flesh is no sooner cut or bruised than the body begins to secrete a substance that tends to heal the cut, or throw off the damaged flesh preparatory to the healing process. So it has been in our National history during the past four or five years. A gigantic war has been crippling the energies of the people, and taxing to its utmost all the resources of the nation. The country has been torn, and although not prostrate, trembles under its exertion, and bleeds under its wounds. The rock opens its bosom, and rivers of oil are poured forth in the very time of our sore visitation. To the reflecting eye God's hand is as visible here as it was when Moses smote the rock in the wilderness to save Israel from dying with thirst. The rod of the Hebrew prophet smote the rock, yet it was Providence that furnished the supply; and now although multitudes of eager men are smiting the rocks, many of them careless of the operations of his Providence, yet it is the same unseen, beneficent hand that brings the oil from its secret cavities, that brought the water to Israel in the time of sore need. And this oil in its proceeds finds its way, by numerous channels, into the public treasury. Sometimes it is in the form of direct tax upon the oil itself, sometimes in tax on incomes, and very largely in stamps, that are used in all the transactions connected with the business.

The tax on refined oil is at present eight dollars per barrel, or twenty cents per gallon. Some of our refiners are paying, when in full operation, from twelve to fifteen hundred dollars per day, bringing an aggregate of about one million of dollars into the national treasury. The present tax on crude oil will yield some two millions more. Transactions in leases and purchases are, and have been, very heavy in Venango county. The business in the registry office has become enormous, occupying generally some nine or ten clerks, and these scarcely able to keep abreast of the business. On some of these, stamps to the amount of twelve hundred dollars have been placed, and on smaller transactions innumerable, that never resulted in any thing to those engaged in them, such as articles of agreement, "refusals," and such like, stamps of greater or less value have been placed.

In many portions of the oil territory, as well as in other places where fortunes have been amassed through the oil business, the income tax has been enormous even under the modest returns that are too often rendered to the officials appointed to receive them. And these taxes are increasing rather than diminishing. Nor are they particularly burdensome to any class of the community. As far as the tax on refined oil is concerned, although it comes from the consumer, yet the article with the tax is more satisfactory and economical than any other illuminator that has been tested. Stamp taxes are almost invariably paid by those who are able to pay them without oppression, and income taxes belong only to those who have an income enabling them to bear their part in the common burden.

We cannot well conceive how the affairs of the Government could have been carried on without this treasure in the rock; but we may admire the medium of the great Provider, in laying it by against our time of need, and bringing it forth just at the proper time.

And since these troublous days are over, and the sword is sheathed to rust away in its scabbard—since the storm has ceased, the gale quieted, and we are called upon to take an observation, and compute our latitude and longitude, this very trade that is so flourishing in our midst may be the means of guiding us safely into port. We have a national debt that is already enormous—a debt that may well frighten even cool heads and well-balanced nerves, as it is contemplated in the future. But with this trade growing strong in our midst, drawing our resources from the earth with comparatively little expense, and sending it to the four quarters of the globe, and to the islands of the ocean, and thus laying contributions upon the nations of the earth, we may well expect to pay off our national debt, and still leave the balance of trade in our favor.

Here, then, is the provision that Providence has made for our time of great need. Here is the exertion that kindly nature is putting forth to heal our wounds and to restore the injuries that we are receiving in this the time of our nation's awful peril.

The history of the oil region would not be completed without a brief reference to the great flood of March, 1865. This deluge, unexampled in modern times, was attended by great disaster and loss of property through-

25

out the entire oil country. It was the result of a rapid melting of snow, accompanied by a heavy rain. The snow had fallen in considerable quantities at the head waters of the Allegheny and French creek, as well as of the smaller tributaries that pertain to these streams. On the 16th of March the rain poured down in torrents nearly the whole day. It was not the ordinary, gradual rain of the spring time, but seemed at times as though the clouds were rent asunder, and the water was precipitated in sheets upon the earth. The waters of French creek and the Allegheny began to rise on the afternoon of that day, but no particular damage was anticipated. But the floods arose during the night with frightful rapidity, and families near the streams found their dwellings invaded, and were obliged to seek a shelter higher up the banks.

On the morning of the 17th the view of French creek and the Allegheny was absolutely frightful. The current was sweeping wildly beyond its usual banks, bearing the evidence of devastation and ruin upon its bosom. Especially was this the case on the Allegheny. It seemed as though the entire oil region must be a scene of ruin. Houses, engine shanties, derricks, oil tanks, bull-wheels, barrels, all in dire confusion, where whirling down the current. Lumber, staves, farming and teaming utensils, all assisted in swelling the mighty tide of destruction. Some fifty houses passed down in the course of the forenoon, some of them rude cabins, and others with shingled roofs. The whole bosom of the river was thickly dotted with oil barrels, some of them filled, but more empty, during the space of twelve or fifteen hours. Even the heavy portable engines were removed from their position, forced into the river, and swept downwards by the resistless current. A few of the houses were swept away during the night, and the tenants were awaked to find themselves tossed about, the sport of the wild waters. These were generally removed by means of boats, although there was some loss of life.

The lower bridge at Franklin, that spans French creek, was lifted from its piers in two sections, and borne down to be dashed against the suspension bridge across the Allegheny. The recoil was dreadful, yet, although the frame and roof of the floating structure were dashed to fragments, the damage to the suspension bridge was but trifling. The Oil Creek bridge came down almost entire, and passed under the suspension without interference. Sugar Creek bridge also came down early in the day or morning.

The entire valleys of many of the streams were swept clear of everything pertaining to oil operations—engines, derricks, tanks filled with oil, barrels empty and filled, houses, offices, all were completely swept away, leaving scarcely a trace behind.

Oil City was completely inundated. The principal street was covered with water to the depth of several feet. Families retreated up the steep bank, and fires were kindled to secure comfort during the night and part of the day.

Franklin did not suffer so much, being situated on higher ground. Still, the street nearest to the creek and river was completely inundated, and some families were removed from their dwellings in boats.

The probabilities are, that such a flood was never known in these valleys before. Citizens who have resided here nearly seventy years testify that during that period no such flood has been experienced, that of 1806 not reaching to anything like the dimensions of the present.

The railroad was swept away, and the damage sustained in the oil operations of the county can only be computed when the flood has completely subsided, and a careful survey taken of the field. When fully summed up it will reach the amount of many millions of dollars. Besides the actual loss of oil and machinery, the probabilities are that the wells themselves have suffered severely from filling up with sand and mud, causing great delay in resuming operations, and, probably, fatal damage through the filling up of the smaller oil veins.

There is a lull in the rush and hurry of business, yet it will be but temporary, and, perhaps, the activity will increase from the temporary suspension. Still gloom rests upon many hearts as they see the labor of months swept away in a single night. Yet such an inundation may not occur again during the next century, and the probabilities are, that business will go on as before on the islands and low grounds, wherever a site can be found or an oil well located, and take the risks of the rain and snow and floods.

# CHAPTER XXIV.

## CONCLUDING THOUGHTS.

In reviewing the great blessings that have come to us in these last days, we should remember that the treasures of the earth belong to the Lord. Even the small things of earth are great in His sight, who takes in all things at a glance. The same hand that decks the heavens with their gorgeous jewelry, paints the wing of the little insect that is crushed beneath the tread of the wayfarer, and that at best runs its brief race in a day. The same ear listens alike to the sparrow's morning song, and to the mighty peal of angelic harps that is eternally echoing from the "sea of glass," spread out around the eternal throne. So the same hand that at first scooped out the channel for the ocean's bed, and reared up the mighty, rock-ribbed mountains of earth, made the earth's bosom the repository for the minerals necessary to man's comfort, and cut out through secret channels in the living rock a pathway for the oil, that he might have wherewith to supply the demands of science, of trade, of utility, and even of luxury. The same wisdom is seen in the small things as in the more important, as we estimate them. If the firmament showeth His handiwork, we see it in the germination of the grain, in the budding of the blossom, in which is wrapped up the embryo fruit, in the opening of the fountains of water, in the storing up the minerals in their mines, and in burying in the flinty rock the rich wealth of oil.

These gifts should be enjoyed with gratitude. If wealth has come to us almost unsought, the great Giver should be remembered with gratitude. A higher authority than that of earth has assured us that "every good gift and every perfect gift is from above, and cometh down from the Father of Lights, with whom there is no variableness, neither shadow of turning." Neither human foresight, nor strong muscle, nor golden coins, nor the perfection of machinery can claim the credit of this vast wealth. Wisdom and power placed it in the rock, and neither man's reason, nor wisdom, nor cunning could reach it until the appointed time had arrived. If it is given freely, with little labor, the grounds of gratitude are all the stronger that should influence the hearts of the recipients of this rich bounty. It is mainly because we will not hearken to Nature's persuasive voice within us, but stifle its low, sweet accents, that we take God's gifts as a matter of course, and enjoy them thankless, as though our own hand had prepared them, and our own heart devised them. If we would but listen to Nature's teaching, and to God's earnest voice, every whisper that Nature utters, and every precept of God's word, would impress upon us lessons of gratitude and thankfulness.

These gifts should induce a liberal spirit. Men should consider themselves but stewards of Heaven's rich bounty, and the almoners of its gifts. If the great Benefactor intrusts us with much, He will require much at our hands. And it is a most blessed thing to be able to relieve want, to urge onward the great enterprises of the day, and to add our portion to the great Spiritual Temple that the Eternal God is erecting, and that he will, ere long, make beautiful in the earth. Even in the matter of common benevolence "it is more blessed to give than to receive."

> " The quality of mercy is not strained:
> It droppeth as the gentle rain from heaven
> Upon the place beneath; it is twice bless'd;
> It blesses him that gives and him that takes;
> 'Tis mightiest in the mightiest; it becomes
> The throned monarch better than his crown."

# INDEX.

THE END.

www.ingramcontent.com/pod-product-compliance
Lightning Source LLC
LaVergne TN
LVHW020239110826
845151LV00003B/970
*9781425529529*